FREE Test Taking Tips DVD Offer

To help us better serve you, we have developed a Test Taking Tips DVD that we would like to give you for FREE. **This DVD covers world-class test taking tips that you can use to be even more successful when you are taking your test.**

All that we ask is that you email us your feedback about your study guide. Please let us know what you thought about it – whether that is good, bad or indifferent.

To get your **FREE Test Taking Tips DVD**, email freedvd@studyguideteam.com with “FREE DVD” in the subject line and the following information in the body of the email:

a. The title of your study guide.

b. Your product rating on a scale of 1-5, with 5 being the highest rating.

c. Your feedback about the study guide. What did you think of it?

d. Your full name and shipping address to send your free DVD.

If you have any questions or concerns, please don’t hesitate to contact us at freedvd@studyguideteam.com.

Thanks again!

ASVAB Study Guide 2021-2022 Pocket Book

ASVAB Test Prep and Practice Exam Questions for the Armed Services Vocational Aptitude Battery [2nd Edition]

TPB Publishing

Written and edited by TPB Publishing.

ISBN 13: 9781628458015
ISBN 10: 1628458011

Table of Contents

Quick Overview

As you draw closer to taking your exam, effective preparation becomes more and more important. Thankfully, you have this study guide to help you get ready.

A large part of the guide is devoted to showing you what content to expect on the exam and to helping you better understand that content. Near the end of this guide is a practice test so that you can see how well you have grasped the content. Then, answer explanations are provided so that you can understand why you missed certain questions.

Once the exam is complete, take some time to relax. Even if you feel that you need to take the exam again, you will be well served by some down time before you begin studying again. It's often easier to convince yourself to study if you know that it will come with a reward!

Test-Taking Strategies

1. Predicting the Answer

When you feel confident in your preparation for a multiple-choice test, try predicting the answer before reading the answer choices. This is especially useful on questions that test objective factual knowledge or that ask you to fill in a blank. By predicting the answer before reading the available choices, you eliminate the possibility that you will be distracted or led astray by an incorrect answer choice.

2. Reading the Whole Question

Too often, test takers scan a multiple-choice question, recognize a few familiar words, and immediately jump to the answer choices. Test authors are aware of this common impatience, and they will sometimes prey upon it.

3. Looking for Wrong Answers

Long and complicated multiple-choice questions can be intimidating. One way to simplify a difficult multiple-choice question is to eliminate all of the answer choices that are clearly wrong. In most sets of answers, there will be at least one selection that can be dismissed right away.

4. Don't Overanalyze

Anxious test takers often overanalyze questions. When you are nervous, your brain will often run wild, causing you to make associations and discover clues that don't actually exist. If you feel that this may be a problem for you, do whatever you can to slow down during the test.

5. Your First Instinct

Many people struggle with multiple-choice tests because they overthink the questions. If you have studied sufficiently for the test, you should be prepared to trust your first instinct once you have carefully and completely read the question and all of the answer choices.

6. Subtle Negatives

One of the oldest tricks in the multiple-choice test writer's book is to subtly reverse the meaning of a question with a word like *not* or *except*. If you are not paying attention to each word in the question, you can easily be led astray by this trick.

Introduction to the ASVAB Test

Function of the Armed Services Vocational Aptitude Battery (ASVAB) Test

The **Armed Services Vocational Aptitude Battery** (ASVAB) measures developed abilities and helps to predict academic and occupational success in the military. Once a military applicant takes the ASVAB, military personnel use the score to assign an appropriate job in the military. The test also assesses whether an applicant is qualified to enlist in the military.

The test can be administered via paper and pencil (P&P-ASVAB) or via computer. The computer test is an adaptive test, which means the test adapts to an individual's ability level based on the answer to the previous question.

Test Administration

The ASVAB is administered annually and is taken by over one million military applicants. Most applicants are high school or post-secondary students. Testing is conducted at a Military Entrance Processing Station (MEPS), or if this site is not available, the ASVAB can be taken at a satellite location known as a Military

Entrance Test (MET) site. The computer-based test takes about one-and-a-half hours for an average person to complete, while the paper and pencil version takes about three hours to complete. If a test taker needs to retake the test, he or she must wait at least a month before retaking. After the second retest, a test taker must wait six months to retest again. Scores may be used for enlistment purposes for up to two years after the date of testing.

Test Format

The paper and pencil test has 225 questions that are broken up into sections with time limits equaling a total of 149 minutes. There is no penalty for guessing on the P&P- ASVAB, and in fact, unanswered questions are marked as incorrect. There is a penalty for incorrect answers on the CAT-ASVAB, so it is ideal for a tester to answer as much as they can without randomly guessing. Unanswered questions also receive a penalty. Both versions of the exam contain four domains: Verbal, Math, Science and Technical, and Spatial. The next page has a chart listing the content tests on the ASVAB and the domain they fall under.

The chart lists the tests in the order in which they appear on the exam.

Test	**Domain**
General Science (GS)	Science/Technical
Arithmetic Reasoning (AR)	Math
Word Knowledge (WK)	Verbal
Paragraph Comprehension (PC)	Verbal
Mathematics Knowledge (MK)	Math
Electronics Information (EI)	Science/Technical
Auto Information (AI)	Science/Technical
Shop Information (SI)	Science/Technical
Mechanical Comprehension (MC)	Science/Technical
Assembling Objects (AO)	Spatial

Scoring

ASVAB scores are determined using an **Item Response Theory** (IRT) model, which tailors the test questions to the ability level of the test taker. After the final ability estimate is measured, a standardized score is reached.

General Science

Geology

Geology is the study of the nature and composition of the rocks and materials that make up the Earth, how they were formed, and the physical and chemical processes that have changed Earth over time.

The theory of **plate tectonics** states that the Earth's lithosphere is a collection of variably-sized plates that move and interact with each other on top of the more molten asthenosphere in the mantle below. Scientists theorize that these plates were once part of a supercontinent, called **Pangea**, that existed over 175 million years ago.

Example

Q. Which of the following is a type of boundary between two tectonic plates?

a. Continental
b. Oceanic
c. Convergent
d. Fault

Explanation

Answer. C: Convergent plate boundaries occur where two tectonic plates collide together. The denser oceanic plate will drop below the continental plate in a process called subduction.

Geography

Geography is the study of the Earth's layout and features. Vertical lines, also called **meridians**, measure east/west distance and are described as **longitudinal**. Horizontal lines measure north/south distance and are described as **latitudinal**. he central latitude line is the **equator**—the line between the Northern and Southern Hemispheres.

Example
Q. Volcanic activity can occur in which of the following?

a. Convergent boundaries
b. Divergent boundaries
c. The middle of a tectonic plate
d. All of the above

Explanation
Answer. D: Volcanic activity can occur at both fault lines and within the area of a tectonic plate at areas called hot spots. Volcanic activity is more common at fault lines because of cracks that allow the mantle's magma to more easily escape to the surface.

Weather, Atmosphere, and the Water Cycle

The study of the Earth's weather, atmosphere, and water cycle is called **meteorology**.

Weather is a state of the atmosphere at a given place and time based on conditions such as air pressure, temperature, and moisture. Weather includes conditions like clouds, storms, temperature, tornadoes, hurricanes, and blizzards.

The **atmosphere** is a layer of gas particles floating in space. The atmospheric levels are created by gravity and its pull on those particles. The Earth's atmosphere is mostly comprised of nitrogen and oxygen (78% and 21%, respectively) along with significantly lower amounts of other gases including 1% argon and 0.039% carbon dioxide.

The **water cycle** is another factor that drives weather. It is the movement of water above, within, and on the surface of the Earth. During any phase of the cycle, water can exist in any of its three phases: liquid, ice, and vapor.

Example

Q. Where is most of the Earth's weather generated?

a. The troposphere
b. The ionosphere
c. The thermosphere
d. The stratosphere

Explanation

Answer. A: Technically, the troposphere is a layer of the atmosphere where the majority of the activity that creates weather conditions experienced on Earth occurs. The ozone layer is in the stratosphere; this is also where airplanes fly.

Astronomy

Astronomy is the study of celestial bodies, or objects in space, and how they interact with each other. Earth is a celestial body; others include the Sun, moon, other planets, black holes, satellites, asteroids, meteors, comets, stars, and galaxies. From what astronomers have observed, the size of the universe is believed to be 91 billion light years and constantly expanding.

Example

Q. What is the largest planet in our solar system and what is it mostly made of?

a. Saturn, rocks
b. Jupiter, ammonia
c. Jupiter, hydrogen
d. Saturn, helium

Explanation

Answer. C: Jupiter is the largest planet in the solar system, and it is primarily composed of hydrogen and helium. Ammonia is in much lower quantity and usually found as a cloud within Jupiter's atmosphere.

Biology

Biology is the study of living organisms. Every living thing on this planet is made of cells. Cells can be split into two categories: prokaryotic and eukaryotic. **Prokaryotic** cells are comparatively quite small and do not have a nucleus. **Eukaryotic** cells contain a nucleus as well as other membrane-bound organelles, which have different functions within the cell and compartmentalize the cell's materials. It's generally believed that unicellular organisms can either be prokaryotic or eukaryotic, but multicellular organisms are always eukaryotic.

Genetics

Genetics is the study of how living organisms pass down traits to their offspring and future generations. It all starts with **DNA**—the universal language of genetic information. An organism's physical features and cellular instructions are written as **genes**, which are single units of genetic information that are stored in every cell as a set of chromosomes.

Anatomy

Anatomy is the science of the structure and parts of organisms.

The **cardiovascular/circulatory system** consists of the heart, blood, and blood vessels. The heart's job is to pump oxygenated blood to the rest of the body through blood vessels called arteries.

The **digestive system** manages food consumption and processing, from chewing to nutrient absorption and defecation. This system consists of the mouth, throat, esophagus, stomach, small and large intestines, and anus.

The **endocrine system** is the hormonal or glandular system of the body. Organ systems communicate with each another via hormones that travel in the bloodstream. This system is vital for maintaining homeostasis (equilibrium of the body).

The **integumentary system** is the skin, hair, nails, sensory receptors, and sweat and oil glands.

The **immune** and **lymphatic system** work together to fight off and clean out elements that are potentially harmful to the body; they consist of white blood cells, lymph nodes, lymphatic vessels, and select organs such as the spleen.

The **musculoskeletal system** contains the muscles, skeleton, cartilage, tendons, and ligaments. In addition to enabling movement, it protects vital organs, such as the heart and lungs inside the ribcage and the brain inside the skull.

The **nervous system** is the body's brain and nerves, similar to a modern car's onboard computer. It's split into two parts: the **central nervous system** (CNS) and **peripheral nervous system** (PNS).

The **renal/urinary system** consists of the kidneys, ureters, bladder, and urethra.

Consisting of the sex organs, the **reproductive system** is the only system that differs in males and females. Males have a penis, prostate, and scrotum containing the testes. Females have a vagina, cervix, uterus, ovaries, fallopian tubes, and mammary glands.

Comprised of the nose, mouth, throat, trachea, lungs, and diaphragm, the **pulmonary/respiratory system** facilitates the exchange of gases—mainly carbon dioxide and oxygen—in a process called **respiration**.

Example

Q. The somatic nervous system is responsible for which of the following?

a. Breathing
b. Thought
c. Movement
d. Fear

Explanation

Answer. C: The somatic nervous system is the voluntary nervous system, responsible for voluntary movement. It includes nerves that transmit signals from the brain to the muscles of the body. Breathing is controlled by the autonomic nervous system. Thought and fear are complex processes that occur in the brain, which is part of the central nervous system.

Chemistry

Periodic Table

The **periodic table** is a chart of all 118 known elements. The elements are organized according to their quantity of protons, their electron configurations, and their chemical properties. The

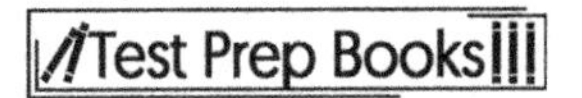

rows are called periods, and the columns are called groups.

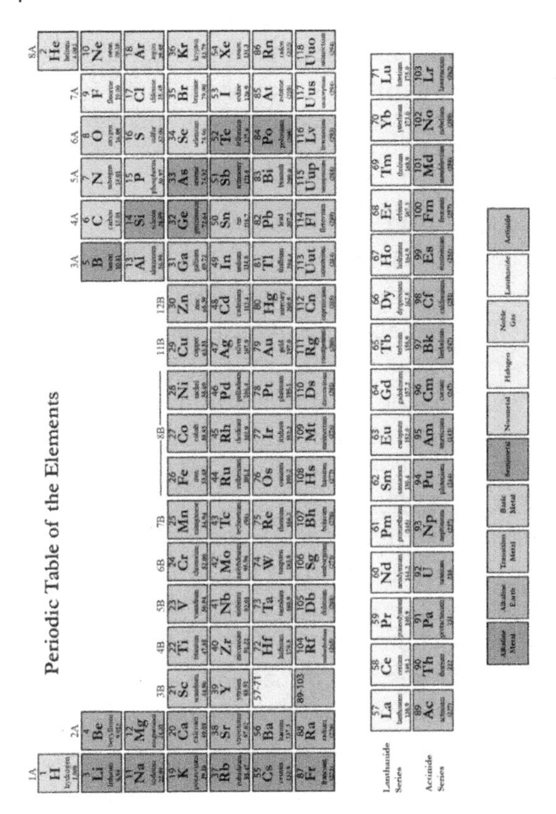

Atomic Structure

Atoms are made up of three subatomic particles: protons, neutrons, and electrons. The **protons** have a positive charge and are located in the nucleus of the atom. **Neutrons** have a neutral charge and are also located in the nucleus. **Electrons** have a negative charge, are the smallest of the three particles, and are located in orbitals that surround the nucleus in a cloud.

Acids and Bases

An **acid** can be described as a substance that increases the concentration of H^+ ions when it is dissolved in water, as a proton donor in a chemical equation, or as an electron-pair acceptor. A **base** can be a substance that increases the concentration of OH^- ions when it is dissolved in water, accepts a proton in a chemical reaction, or is an electron-pair donor.

Chemical Reactions

Chemical reactions are characterized by a chemical change in which the starting substances, or **reactants**, differ from the substances formed, or **products**. Chemical reactions may involve a change in color, the production of gas, the formation of a precipitate, or changes in heat content. The following are the basic types of chemical reactions:

- Decomposition Reaction
- Synthesis Reaction
- Single Displacement Reaction
- Double Displacement Reaction

- Combustion Reaction

Example

Q. What type of chemical reaction produces a salt?

a. Oxidation reaction
b. Neutralization reaction
c. Synthesis reaction
d. Decomposition reaction

Explanation

Answer. B: A neutralization reaction produces a salt. A solid produced during a reaction is called **precipitation.** A precipitation reaction can be used for removing a salt (an ionic compound that results from a neutralization reaction) from a solvent such as water. For water, this process is called ionization. Therefore, the products of a neutralization reaction (when an acid and base react) are a salt and water. Choice *A*, oxidation reaction, involves the transfer of an electron. Choice *C*, synthesis reaction, involves the joining of two molecules to form a single molecule. Choice *D*, decomposition reaction, involves the separation of a molecule into two other molecules.

Physics

Energy

Energy may be defined as the capacity to do work. It can be transferred between objects, comes in a multitude of forms, and can be converted from one form to another. The **Law of Conservation of Energy** states, "Energy can neither be created nor be destroyed."

Mechanics

The description of motion is known as **kinetics**, and the causes of motion are known as **dynamics**. Motion in one dimension is a **scalar** quantity. It consists of one measurement such as length, speed, or time. Motion in two dimensions is a **vector** quantity. This is a speed with a direction, or **velocity**.

Newton's Three Laws of Motion

Sir Isaac Newton summarized his observations and calculations relating to motion into three concise laws:

First Law of Motion: Inertia

This law states that an object in motion tends to stay in motion or an object at rest tends to stay at rest unless the object is acted upon by an outside force.

Second Law of Motion: $F = ma$

This law states that the force on a given body is the result of the object's mass multiplied by the acceleration acting upon the object. For objects falling on Earth, acceleration is caused by gravitational force ($9.8\ m/s^2$).

Third Law of Motion: Action-Reaction

This law states that for every action there is an equal and opposite reaction.

Example

Q. Circular motion occurs around what?

a. The center of mass
b. The center of matter
c. An elliptical
d. An axis

Explanation

Answer. D: Circular motion occurs around an invisible line around which an object can rotate; this invisible line is called an axis. Choice *A*, center of mass, is the average placement of an object's mass. Choice *B* is not a real term. Choice *C*, elliptical, describes an elongated circle and is not a viable selection.

Arithmetic Reasoning

Basic Operations of Arithmetic

There are four basic operations used with numbers: **addition, subtraction, multiplication**, and **division**.

Exponentiation is a kind of repeated multiplication. 3^4, read "three to the fourth power," means to multiply 3 by itself 4 times: $3 \times 3 \times 3 \times 3 = 81$. The exponent 2 is the exponentiation case used most often in calculations. In that case, it is called squaring the number.

A **fraction** is a number used to express a ratio. It is written as a number *x* over a line with another number *y* underneath: $\frac{x}{y}$. The number on top (*x*) is called the **numerator**, and the number on the bottom (*y*) is called the **denominator**.

Percentages are essentially just fractions out of 100 (the word comes from the Latin meaning "per one hundred") and are written with the % symbol. So, 35% means $\frac{35}{100}$.

Example

Q. A traveler takes an hour to drive to a museum, spends 3 hours and 30 minutes there, and takes half an hour to drive home. What percentage of his or her time was spent driving?

a. 15%
b. 30%
c. 40%
d. 60%

Explanation

Answer. B: The total trip time is 1 + 3.5 + 0.5 = 5 hours. The total time driving is 1 + 0.5 = 1.5 hours. So, the fraction of time spent driving is 1.5/5 or 3/10. To get the percentage, convert this to a fraction out of 100. The numerator and denominator are multiplied by 10, with a result of 30/100. The percentage is the numerator in a fraction out of 100, so 30%.

Basic Geometry Relationships

Perimeter is the measurement of a distance around something. It is found by adding together the lengths of all of the sides of a figure.

Here are the formulas for the perimeter of a square and the perimeter of a rectangle:

- Perimeter of a square: $P = 4 \times s$
- Perimeter of a rectangle: $P = l + l + w + w = 2l + 2w$

In contrast to perimeter, **area** is the space occupied by a defined enclosure, like a field enclosed by a fence. It is measured in square units such as square feet or square miles.

Here are some formulas for the areas of basic planar shapes:

- The area of a rectangle is $l \times w$, where *w* is the width and *l* is the length
- The area of a square is s^2, where *s* is the length of one side (this follows from the formula for rectangles)
- The area of a triangle with base *b* and height *h* is $\frac{1}{2}bh$
- The area of a circle with radius *r* is πr^2

Volume is the measurement of how much space an object occupies. Volume questions will typically ask how much of something is needed to completely fill the object. It is measured in cubic units, such as cubic inches.

Here are some formulas for the volumes of basic three-dimensional geometric figures:

- For a regular prism whose sides are all rectangles, the volume is $l \times w \times h$, where *w* is the width, *l* is the length, and *h* is the height of the prism
- For a cube, which is a prism whose faces are all squares of the same size, the volume is s^3
- The volume of a sphere of radius *r* is given by $\frac{4}{3}\pi r^3$
- The volume of a cylinder whose base has a radius of *r* and which has a height of *h* is given by $\pi r^2 h$

Example

Q. A truck is carrying three cylindrical barrels. Their bases have a diameter of 2 feet, and they have a height of 3 feet. What is the total volume of the three barrels in cubic feet?

a. 3π
b. 9π
c. 12π
d. 15π

Explanation

Answer. B: The formula for the volume of a cylinder is $\pi r^2 h$, where *r* is the radius and *h* is the height. The diameter is twice the radius, so these barrels have a radius of 1 foot. That means each barrel has a volume of $\pi \times 1^2 \times 3 = 3\pi$ cubic feet. Since there are three of them, the total is $3 \times 3\pi = 9\pi$ cubic feet.

Word Knowledge

Defining Words and English Origins

A **word** is a group of letters joined to form a single meaning. Words can be nouns, verbs, adjectives, and adverbs, among others. Words also represent a verb tense of past, present, or future. Words allow for effective communication for commerce, social progress, technical advances, and much more. Simply put, words allow people to understand one another and create meaning in a complex world.

<u>Example</u>

Q. **Wave** most nearly means

a. flourish
b. sink
c. tide
d. stagnant

<u>Explanation</u>

Answer. A: This was a tricky question because some choices connect **wave** to actions involved with water. **Sink** and **tide** are actions associated with the water, but are distinct from an ocean wave, which means wave is not used in context of an ocean wave, but as a verb to wave. To wave is to move around. **Stagnant** is to be still. **Flourish** describes a waving motion, making it the correct match.

Word Formation

Roots and Root Words

Root words are words written in their most basic form, and they carry a clear and distinct meaning. Consider the word *safe*. The root word, *safe*, acts as both a noun and adjective, and stands on its own, carrying a clear and distinct meaning.

The root of a word, however, is not necessarily a part of the word that can stand on its own, although it does carry meaning. Since many English words come from Latin and Greek roots, it's helpful to have a general understanding of roots. Here is a list of common Greek and Latin roots used in the English language:

Root	Definition	Example
ami	love	amiable
ethno	race	ethnological
infra	beneath or below	infrastructure
lun	moon	Lunar

<u>Affixes</u>

Affixes are groups of letters that when added to the beginning or ending of root words, or are attachments within a root or root word itself, can:

- Intensify the word's meaning
- Create a new meaning
- Somewhat alter the existing meaning
- Change the verb tense
- Change the part of speech

There are three types of affixes: prefixes, suffixes, and infixes.

Prefixes

Prefixes are groups of letters attached to the beginning of a root word. *Pre-* refers to coming before, and *fix* refers to attaching to something. Consider the example of the root word *freeze*:

- Freeze: verb – to change from a liquid to solid by lowering the temperature to a freezing state.
- *Anti*freeze: noun – a liquid substance that prevents freezing when added to water, as in a vehicle's radiator.

By adding the prefix *anti-* to the root word *freeze*, the part of speech changed from verb to noun, and completely altered the meaning. *Anti-* as a prefix always creates the opposite in meaning, or the word's antonym.

By having a basic understanding of how prefixes work and what their functions are in a word's meaning, English speakers strengthen their fluency. Here is a list of some common prefixes in the English language, accompanied by their meanings:

Prefix	**Definition**	**Example**
ante-	before	antecedent
ex-	out/from	expel
inter-	between/among	intergalactic
multi-	much/many	multitude

Suffixes

Suffixes are groups of letters attached to the ending of a root or root word. Like prefixes, suffixes can:

- Intensify the word's meaning
- Create a new meaning
- Somewhat alter the existing meaning
- Change the verb tense
- Change the part of speech

Consider the example of the root word *fish* when suffixes are added:

- Fish: noun – a cold-blooded animal that lives completely in water and possesses fins and gills.
- Fishing: noun – I love the sport of fishing.
- Fishing: verb – Are you fishing today?

With the addition of the suffix *-ing*, the meaning of root word *fish* is altered, as is the part of speech.

A verb tense shift is made with the addition of the suffix *-ed*:

- Jump: present tense of to jump as in "I jump."
- Jumped: past tense of to jump as in "I jumped."
- Climb: present tense of to climb as in "We climb."
- Climbed: past tense of to climb as in "We climbed."

Here are a few common suffixes in the English language, along with their meanings:

Suffix	Meaning	Example
-ed	past tense	cooked
-ing	materials, present action	clothing
-ly	in a specific manner	lovely
-ness	a state or quality	brightness

Infixes

Infixes are letters that attach themselves inside the root or root words. They generally appear in the middle of the word and are rare in the English language. Easily recognizable infixes include parent*s*-in-law, passer*s*-by, or cup*s*ful. Notice the ***-s*** is added inside the root word, making the word plural.

Example

Q. The toddler, who had just learned to speak, seemed rather **loquacious**.

a. verbose
b. humorous
c. silent
d. cranky

Explanation

Answer. A: Loquacious reflects the tendency to talk a lot. While someone can be humorous and loquacious,

humorous describes the kind of talk, not the fact that someone talks a lot. **Cranky** has nothing to do with the word, and **silent** is a clear opposite. This leaves **verbose**, meaning using an abundance of words; loquacious is a synonym.

Paragraph Comprehension

Topic Versus Main Idea

A **topic** is the subject of the text; it can usually be described in a one- to two-word phrase and appears in the simplest form. On the other hand, the **main idea** is more detailed and provides the author's central point of the text. It can be expressed through a complete sentence.

Example

Do you want to vacation at a Caribbean island destination? Who wouldn't want a tropical vacation? Visit one of the many Caribbean islands where visitors can swim in crystal blue waters, swim with dolphins, or enjoy family-friendly or adult-only resorts and activities. Every island offers a unique and picturesque vacation destination. Choose from these islands: Aruba, St. Lucia, Barbados, Anguilla, St. John, and so many more. A Caribbean island destination will be the best and most refreshing vacation ever . . . no regrets!

Q. What is the topic of the passage?

a. Caribbean island destinations
b. Tropical vacation
c. Resorts
d. Activities

Explanation

Answer. A: The topic of the passage is Caribbean island destinations. The topic of the passage can be described in a one- or two-word phrase. Remember, when paraphrasing a passage, it is important to include the topic. Paraphrasing is when one puts a passage into his or her own words.

Supporting Details

Supporting details help readers better develop and understand the main idea. Supporting details answer questions like *who, what, where, when, why,* and *how*. Different types of supporting details include examples, facts and statistics, anecdotes, and sensory details.

Example

Do you want to vacation at a Caribbean island destination? Who wouldn't want a tropical vacation? Visit one of the many Caribbean islands where visitors can swim in crystal blue waters, swim with dolphins, or enjoy family-friendly or adult-only resorts and activities. Every island offers a unique and picturesque vacation destination. Choose from these islands: Aruba, St. Lucia, Barbados, Anguilla, St. John, and so many more. A Caribbean island destination will be the best and most refreshing vacation ever . . . no regrets!

Q. What is/are the supporting detail(s) of this passage?

a. Cruising to the Caribbean
b. Local events
c. Family or adult-only resorts and activities
d. All of the above

Explanation

Answer. C: Family or adult-only resorts and activities are supporting details in this passage. Supporting details are details that help readers better understand the main idea. They answer questions such as *who, what, where, when, why*, or *how*. In this question, cruises and local events are not discussed in the passage, whereas family and adult-only resorts and activities support the main idea.

Theme

The **theme** of a text is the central message of the story. The theme can be about a moral or lesson that the author wants to share with the audience. The theme of a text can center around varying subjects such as courage, friendship, love, bravery, facing challenges, or adversity. It often leaves readers with more questions than answers. Authors tend to insinuate certain themes in texts; however, readers are left to interpret the true meaning of the story.

Example
The fairy tale *The Boy Who Cried Wolf* features the tale of a little boy who continued to lie about seeing a wolf. When the little boy actually saw a wolf, no one believed him because of all of the previous lies. The author of this fairy tale does not directly tell readers, "Don't lie because people will question the credibility of the story." The author simply portrays the story of the little boy and presents the moral through the tale.

The theme of a text can center around varying subjects such as courage, friendship, love, bravery, facing challenges, or adversity. It often leaves readers with more questions than answers. Authors tend to insinuate certain themes in texts; however, readers are left to interpret the true meaning of the story.

Purposes for Writing

An author's purpose for a text may be to persuade, inform, entertain, or be descriptive. When an author tries to **persuade** a reader, the reader must be cautious of the intent or argument. If an author writes an informative text, their purpose is to **inform** or educate the reader about a certain topic. Entertaining texts, whether fiction or nonfiction, are meant to captivate readers' attention or to **entertain**. Descriptive texts use adjectives and adverbs to

describe people, places, or things to provide a clear image to the reader throughout the story.

Example

Do you want to vacation at a Caribbean island destination? Who wouldn't want a tropical vacation? Visit one of the many Caribbean islands where visitors can swim in crystal blue waters, swim with dolphins, or enjoy family-friendly or adult-only resorts and activities. Every island offers a unique and picturesque vacation destination. Choose from these islands: Aruba, St. Lucia, Barbados, Anguilla, St. John, and so many more. A Caribbean island destination will be the best and most refreshing vacation ever . . . no regrets!

Q. What is the author's purpose of this passage?

a. Entertain readers
b. Persuade readers
c. Inform readers
d. None of the above

Explanation

Answer. B: The author of the passage is trying to persuade readers to vacation in a Caribbean island destination by providing enticing evidence and a variety of options. The passage even includes the author's opinion. Not only does the author provide many details to support his or her opinion, the author also implies that the reader would almost be "in the wrong" if he or she didn't want to visit a Caribbean island, hence, the author is trying to persuade the reader to visit a Caribbean island.

Point of View

Point of view is the perspective in which authors tell stories. Authors can tell stories in either the first, second, or third person. When authors write in the first person, the character refers to themselves as "I." The pronouns *I* and *we* are used when writing in the first person. If an author writes in second person, they are addressing the reader directly, using the pronoun "you." If an author writes in the third person, the narrator is telling the story from an outside perspective, using pronouns such as *he, she, it,* and *they*.

Q. Which of the following sentences uses second person point of view?

a. I don't want to make plans for the weekend before I see my work schedule.
b. She had to miss the last three yoga classes due to illness.
c. Pluto is no longer considered a planet because it is not gravitationally dominant.
d. Be sure to turn off all of the lights before locking up for the night.

Explanation

Answer. D: Choice *D* directly addresses the reader, so it is in second person point of view. This is an imperative sentence since it issues a command; imperative sentences have an *understood you* as the subject. Choice *A* uses first person pronouns *I* and *my*. Choices B and C are incorrect because they use third person point of view.

Types of Passages

Authors write with different purposes in mind. They use a variety of writing passages to appeal to their chosen audience. There are five types of writing passages:

- Narrative
- Expository
- Descriptive
- Persuasive
- Technical

Narrative writing: When an author writes a narrative, they are telling a story. Narratives develop characters, drive a sequence of events, and deal with conflict.

Expository writing: Expository writing is meant to instruct or inform and usually lacks any kind of persuasive elements. Expository writing includes academic writing or newspaper articles.

Descriptive writing: Descriptive writing is writing that uses imagery and figurative language in order to allow the reader to feel as if they are experiencing the text firsthand.

Persuasive writing: Persuasive writing is used when someone is writing an argument. Authors using persuasive writing are attempting to change the opinions and attitudes of their audience.

Technical writing: Technical writing is a genre that includes how-to manuals, recipes, or data-based information. Technical writing is usually straightforward and has no persuasive elements to it.

<u>Example</u>
Q. Which type of writing is meant to instruct of inform?

a. Descriptive writing
b. Narrative writing
c. Persuasive writing
d. Expository writing

<u>Explanation</u>
Answer. D: Expository writing is meant to instruct or inform. Expository writing does not tell a story,

persuade the author, or rely only on descriptive language to get its point across.

Opinions, Facts, and Fallacies

Facts Versus Opinions

A **fact** is a piece of information that is true. It can either prove or disprove claims or arguments presented in texts. Facts cannot be changed or altered.

An **opinion** is a belief or view formed about something that is not necessarily based on the truth. Opinions often express authors' personal feelings about a subject and use words like *believe, think,* or *feel.*

Bias and Stereotypes

A **bias** is when someone demonstrates a prejudice in favor of or against something or someone in an unfair manner. When an author is biased in his or her writing, readers should be skeptical despite the fact that the author's bias may be correct.

A **stereotype** shows favoritism or opposition but toward a specific group or place. Stereotypes create an oversimplified or overgeneralized idea about a certain group, person, or place.

<u>Example</u>

Q. Which of the following is an opinion, rather than historical fact?

a. Leif Erikson was definitely the son of Erik the Red; however, historians debate the year of his birth.
b. Leif Erikson's crew called the land Vinland since it was plentiful with grapes.
c. Leif Erikson deserves more credit for his contributions in exploring the New World.
d. Leif Erikson explored the Americas nearly five hundred years before Christopher Columbus.

<u>Explanation</u>

Answer. C: Choice *C* is the correct answer; it is the author's opinion that Erikson deserves more credit, not a verifiable fact.

Organization of the Text

The **structure of the text** is how authors organize information in their writing. There are various types of patterns in which authors can organize texts, some of which include problem and solution, cause and effect, chronological order, and compare and contrast. These four text structures are described below.

Problem and Solution

One way authors can organize their text is by following a **problem and solution** pattern. This type of structure may present the problem first without offering an immediately clear solution. This pattern may also offer the solution first and then hint at the problem throughout the text. Some texts offer multiple solutions to the same problem.

Cause and Effect

Cause and effect is one of the more common ways that authors organize texts. In a cause and effect text, the author explains what caused something to happen. Persuasive and expository writing models frequently use a cause and effect organizational pattern as well.

Chronological Order

When using a **chronological order** organizational pattern, authors simply state information in the order in which it occurs.

Compare and Contrast

The **compare and contrast** organizational text pattern explores the differences and similarities of two or more objects.

Example

Rain, rain, go away, come again another day. Even though the rain can put a damper on the day, it can be helpful and fun, too. For one, the rain helps plants grow. Without rain, grass, flowers, and trees would be deprived of vital nutrients they need to develop. Not only does the rain help plants grow, on days where there are brief spurts or sunshine, rainbows can appear. The rain reflects and refracts the light, creating beautiful rainbows in the sky. Finally, puddle jumping is another fun activity that can be done in or after the rain. Therefore, the rain can be helpful and fun.

Q. What is the *cause* in this passage?

a. Plants growing
b. Rainbows
c. Puddle jumping
d. Rain

Explanation

Answer. D: Rain is the *cause* in this passage because it is why something happened. The effects are plants growing, rainbows, and puddle jumping.

Math Knowledge

Numbers and Their Classification

- **Whole numbers** are the basic counting numbers.
- **Integers** are whole numbers together with the negative versions of them.
- A **factor** of an integer is a positive integer that divides it evenly. For example, 2 is a factor of 8.
- An **even number** is an integer for which 2 is a factor. If 2 is not a factor of an integer, then it is said to be an **odd number**.
- A **common factor** of multiple integers is a number that is a factor for each of them.
- A **prime number** is a whole number larger than 1 whose only factors are 1 and itself. For example, 2, 3, and 5 are prime numbers.
- A **composite number** is a whole number that is not prime. For example, 4 and 8 are not prime numbers because they have the factor 2.
- A **multiple** of a whole number is any number that can be obtained by multiplying the whole number by another whole number.

- A **common multiple** of a set of whole numbers is a number that is a multiple of all of the numbers in the set. For example, 4 has multiples of 4, 8, 12, 16, 20, 24, etc.
- **Rational numbers** are numbers that can be written as a fraction whose numerator and denominator are both integers.
- A **decimal** is a number that uses a **decimal point** to show that a part of the number is less than 1. For example, 10.3 = 10 + $\frac{3}{10} = 10.3$.
- The **decimal place** is how far to the right of the decimal point a digit appears.
- **Real numbers** include rational numbers as well as any other number that can be expressed using decimals.
- The number system usually used is the **decimal system**, which uses the numerals 0, 1, 2, 3, 4, 5, 6, 7, 8, and 9.

Example

Q. The factors of $2x^2 - 8$ are:

a. $2(x-2)(x-2)$
b. $2(x^2+4)$
c. $2(x+2)(x+2)$
d. $2(x+2)(x-2)$

Explanation

Answer. D: The easiest way to approach this problem is to factor out a 2 from each term:

$$2x^2 - 8 = 2(x^2 - 4)$$

The formula $x^2 - y^2 = (x + y)(x - y)$ can be used to factor $x^2 - 4 = x^2 - 2^2 = (x + 2)(x - 2)$.

So:

$$2(x^2 - 4) = 2(x + 2)(x - 2)$$

Operations

An **exponent** is written as x^y, but is read "*x* to the *y*," or "*x* to the power of *y*," and indicates how many times to multiply *x* by itself. In this expression, *x* is called the **base**, and *y* is called the **exponent**. The case where the exponent is 2 is called **squaring**, and the case where the power is 3 is called **cubing**. So $4^3 = 4 \times 4 \times 4 = 64$.

A **root** is a number that, when raised to some power, gives the number inside. It is written as $\sqrt[n]{x}$. This is called the *n*-th root of *x*, and indicates the number that, when raised to the power of *n*, gives us *x*.

Parentheses are used to show the order in which operations are to be performed in an expression with multiple operations and terms. The rule to follow is to do the operations inside the parentheses first. However, not every operation will be marked off with parentheses.

The order in which the operations are performed is as follows:

- **Parentheses**
- **Exponents**
- **Multiplication**
- **Division**
- **Addition**
- **Subtraction**

Some students find it helpful to memorize this order by using a mnemonic such as **PEMDAS**: *Please Excuse My Dear Aunt Sally.*

Scientific notation is a way to write very large or very small numbers in the form $x \times 10^n$, where *x* is a number between 1 and 10. To get the number in scientific notation, the decimal point is moved *n* places to the right if *n* is positive (filling in zeroes if needed), and *n* places to the left of *n* is negative.

Example

Q. Simplify $(2x-3)(4x+2)$

a. $8x^2-8x-6$

b. $6x^2+8x-5$

c. $-4x^2-8x-1$

d. $4x^2-4x-6$

Explanation

Answer. A: To solve this problem, one has to multiply each of the terms in the first parentheses and then multiply each of the terms in the second parentheses:

$$\begin{aligned}(2x-3)(4x+2) \\ &= 2x(4x)+2x(2)-3(4x)-3(2) \\ &= 8x^2+4x-12x-6 \\ &= 8x^2-8x-6\end{aligned}$$

Systems of Equations

A **linear system** of equations with two variables and two equations is a system with variables *x* and *y* and equations that can be simplified to yield $ax+by=c, dx+ey=f$. There are two ways to solve such a system. The first is to solve for one variable in terms of the other and substitute it into the other equation. For example, from the first equation, $by=c-ax$, that means $y=\frac{c-ax}{b}$. $\frac{c-ax}{b}$ can be substituted for *y* in the second equation. This approach is called solving by **substitution**.

The other possibility is to multiply one of the equations on both sides by some constant, and then add the result to the other equation so that it eliminates one variable. This approach is called solving by **elimination**.

Example

Q. Solve for *x* and *y*, given $3x + 2y = 8, -x + 3y = 1$.

a. $x = 2, y = 1$
b. $x = 1, y = 2$
c. $x = -1, y = 6$
d. $x = 3, y = 1$

Explanation

Answer. A: From the second equation, the first step is to add *x* to both sides and subtract 1 from both sides:

$$-x + 3y + x - 1 = 1 + x - 1$$

with the result of:

$$3y - 1 = x$$

Then, this is substituted into the first equation, yielding:

$$3(3y - 1) + 2y = 8$$

$$9y - 3 + 2y = 8$$

$$11y = 11$$

$$y = 1$$

Then, this value is plugged into:

$$3y - 1 = x, \text{ so } 3(1) - 1 = x \text{ or } x = 2, y = 1$$

Geometry and Angles

An **angle** describes the separation or gap between two lines meeting at a single point. It is written with the symbol ∠. The point where the lines or line segments meet is called the **vertex** of the angle. If the angle is formed by lines that cross one another, the vertex is the point where they cross.

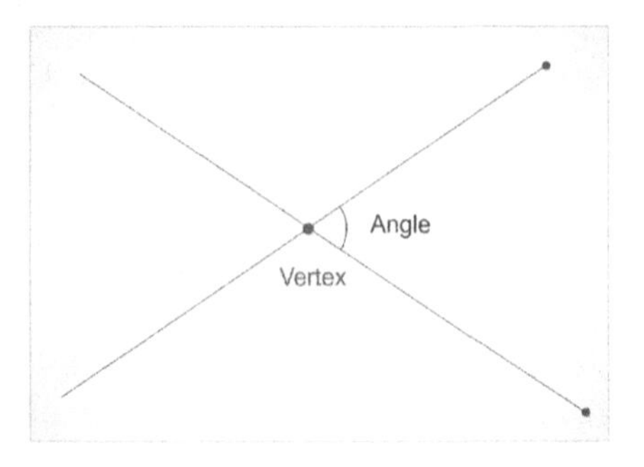

A **right angle** is 90°. An **acute angle** is an angle that is less than 90°. An **obtuse angle** is an angle that is greater than 90° but less than 180°.

An angle of 180° is called a **straight angle**. This is really when two line segments meet at a point, but go in opposite directions, so that they form a single line segment, extending in opposite directions.

A **full angle** is 360°. It is equivalent to spinning all of the way around from facing one direction back to that same direction. A full circle is considered 360°.

If the sum of two angles is 90°, the angles are **complementary**.

If the sum of two angles is 180°, the angles are **supplementary**.

When two lines intersect, the pairs of angles they form are always supplementary. The two angles marked below are supplementary:

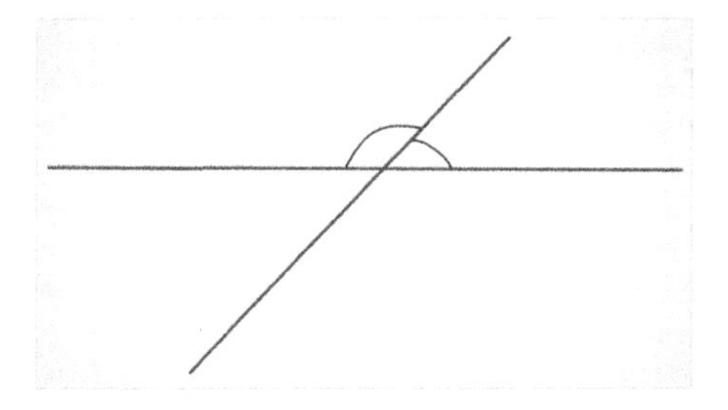

When two supplementary angles are next to one another or "adjacent" in this way, they always give rise to a straight line.

A **triangle** is a geometric shape formed by 3 line segments whose endpoints agree.

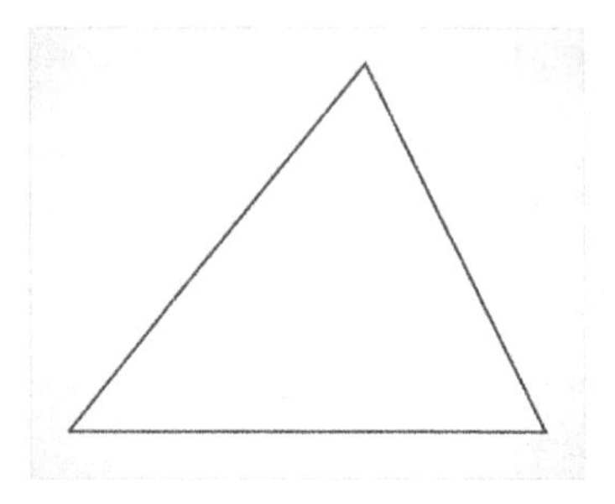

Triangle

The three angles inside the triangle are called **interior angles** and add to 180°. Triangles can be classified by the kinds of angles they have and the lengths of their sides.

An acute triangle is a triangle whose angles are all less than 90°.

If one of the angles in a triangle is 90°, then the triangle is called a **right triangle**.

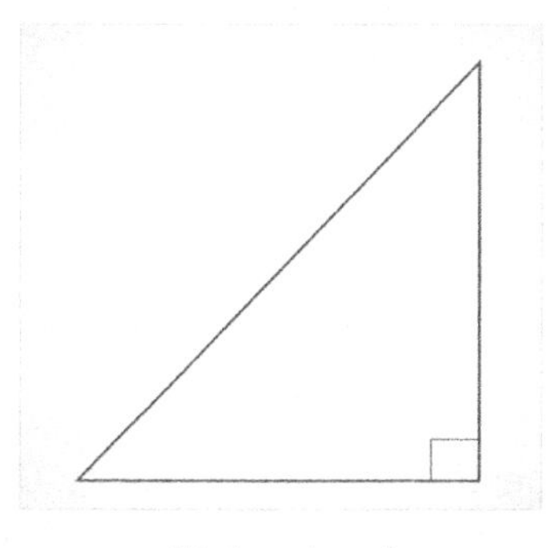

Right triangle

If one of the angles is larger than 90°, then the triangle is called **obtuse**.

An **isosceles triangle** has two sides of equal length. Equivalently, it has two angles that are the same. It can be an acute, right, or obtuse triangle.

A **scalene triangle** has three sides of different length. It also has three unequal angles. An **equilateral triangle** is a triangle whose three sides are the same length. Accordingly, its three angles are equal, and are 60°.

Consider the following triangle with the lengths of the sides labeled as *A*, *B*, *C*.

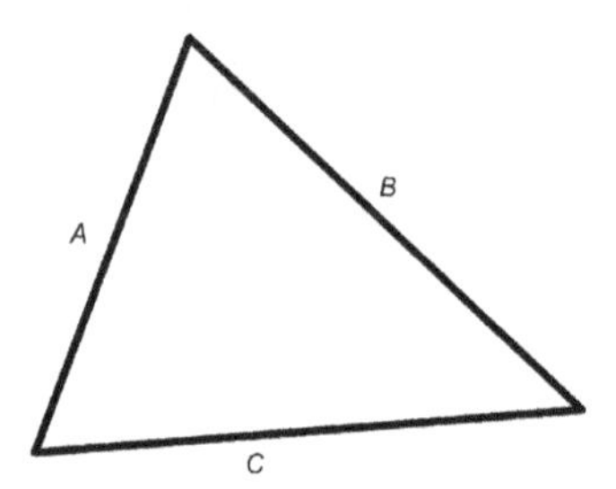

For any triangle, the **Triangle Inequality Theorem** says that the following holds true: $A + B > C, A + C > B, B + C > A$. In addition, the sum of two angles must be less than 180°.

If two triangles have angles that agree with one another, that is, the angles of the first triangle are equal to the angles of the second triangle, then the triangles are called **similar**. Similar triangles look the same, but one can be a "magnification" of the other.

Two triangles with sides that are the same length must also be similar. In this case, such triangles are called **congruent**. Congruent triangles have the same angles and lengths, even if they are rotated relative to one another.

Example

Q. A triangle is to have a base 1/3 as long as its height. Its area must be 6 square feet. How long will its base be?

a. 1 foot
b. 1.5 feet
c. 2 feet
d. 2.5 feet

Explanation

Answer. C: The formula for the area of a triangle with base b and height h is $\frac{1}{2}bh$. In this problem, the base is one-third the height, or $b = \frac{1}{3}h$ or equivalently $h = 3b$. Using the formula for a triangle, this becomes $\frac{1}{2}b(3b) = \frac{3}{2}b^2$. The problem states that this has to equal 6. So, $\frac{3}{2}b^2 = 6, b^2 = 4, b = \pm 2$. However, lengths are positive, so the base must be 2 feet long.

Electronics Information

Electric Charge

Electricity is a form of energy, like heat or movement, that can be harnessed to perform useful work. Electrical energy results from the electric force that exists between atoms and molecules with electrical charge, which is associated with the atomic structure of those substances.

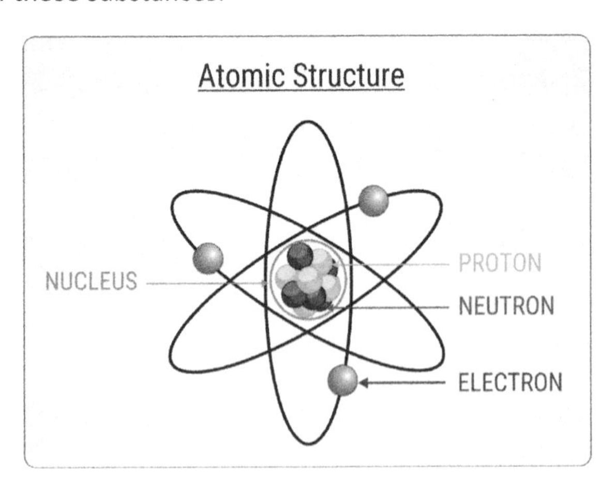

Example

Q. The energy from electricity results from which of the following?

a. The atomic structure of matter
b. The ability to do work
c. The neutrons in an atom
d. Conductive materials like metals

Explanation

Answer. A: The physical structure of the atoms that compose matter lends itself to the production of electricity. The arrangement of the subatomic particles and the associated charges—mainly the negatively charged electrons in the cloud—are associated with the ability to create an electric current, which can be harnessed to do work.

Current

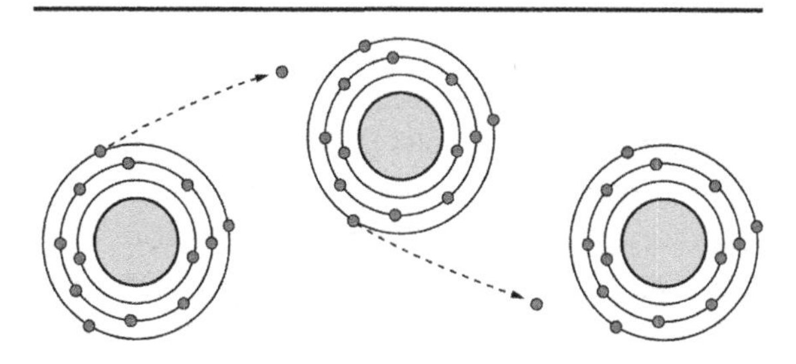

Current is the rate at which the electrical charge flows through a conductive material. It is measured in amperes (A), with each ampere equal to approximately 6.24×10^{18} electrons per second.

Example

Electric current carries energy much like pipes carry water. Water tanks are often elevated and have various pipes to transport water to houses. The purpose behind elevating the tank is to create pressure in the pipes carrying the water. This pressure results in water flow, which can be equated to voltage (akin to pressure) that pushes electrons (the "water") through a circuit. With the water pipes, when the faucet is closed, water does not flow through the pipes, but water flows faster and faster the more the faucet is turned on. The rate of water flow is analogous to current—the rate of electron flow through a circuit.

Voltage

Voltage is the push, or potential, behind electrical work. It is measured in volts (V) and can be thought of as the electromotive potential. Voltage causes current, such that if there is a closed path for electrons and a voltage, current will flow. If there is a suitable path for electrons but no voltage, or voltage in the absence of a viable path, current will cease.

Example

Q. What is the voltage lost at the 25-ohm resistor?

a. 0.22 amps

b. 0 amps

c. 7.25 amps

d. 5.4 volts

Explanation

Answer. A: The voltage lost can be calculated by substituting known values into iterations of Ohm's Law. $V = I \times R$ = (0.29 amps) × (18.75 ohms) = 5.4 volts. Then, this is substituted into the following equation:

$I = V/R$ = 5.4 V/25 ohms = 0.22 amps

Resistance

Electrical resistance, measured in ohms (Ω), is the amount of pressure inhibiting the flow of electrical current. Like friction, which slows the rate of movement, resistance dissipates energy and reduces the rate of flow or the movement of current. The amount of resistance that a given object contributes to a circuit depends on the properties of the object, particularly the material. Materials that are inherently more resistant inhibit the ease at which the electrons in the material's atoms can be displaced.

Example

Q. Which of the following statements is true?

a. Conductors have high resistance because the electrons are easily detached from the atoms.
b. Conductors have high resistance because the electrons are not easily detached from the atoms.
c. Conductors have low resistance because the electrons are easily detached from the atoms.
d. Conductors have low resistance because the electrons are not easily detached from the atoms.

Explanation

Answer. C: Electric conductors tend to be metals, such as silver, copper, and aluminum. They "conduct" electricity, which means that they help the electric current flow easily. This is largely because metals have low resistance because the electrons are easily detached from these atoms, so they are then free to jump from atom to atom and create a current.

Basic Circuits

A **circuit** is a closed loop through which current can flow. A simple circuit contains a voltage source and a resistor. The current flows from the positive side of the voltage source through the resistor to the negative side of the voltage source. Note that if the switch is open or there is some other disconnected wire or break in continuity in the circuit, there will be no electromotive force; the circuit must be a closed loop to create a net flow of electrons from the voltage source through the wires and system.

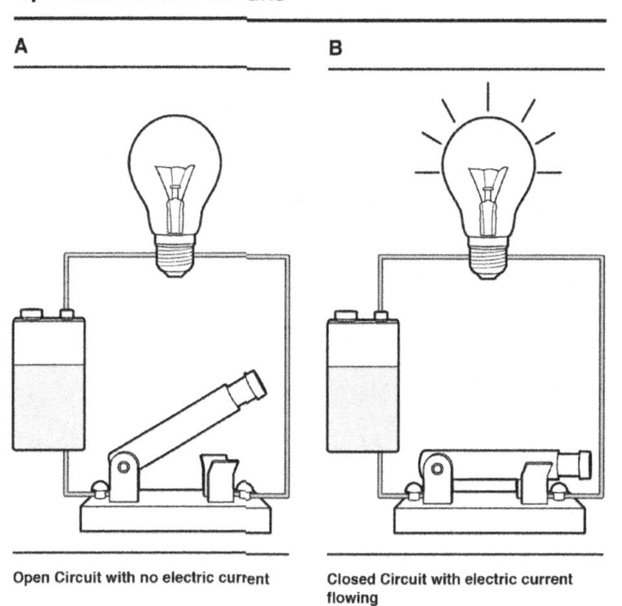

Open Circuit with no electric current

Closed Circuit with electric current flowing

Example

Q. In the following circuit, what is the total resistance across the two terminals (A and B)?

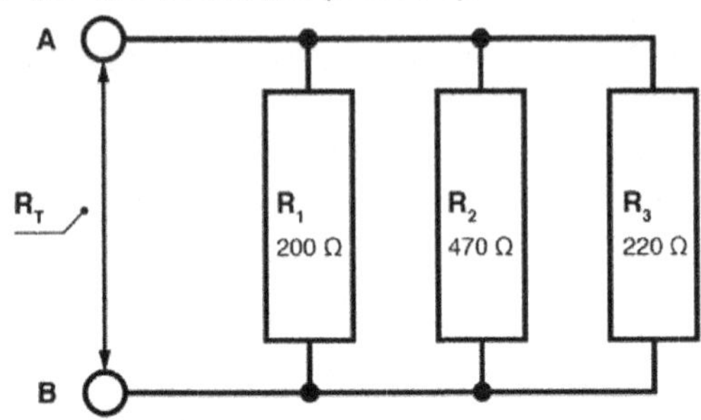

a. 85.67 ohms
b. 790 ohms
c. 0.0117 ohms
d. 200 ohms

Explanation

Answer. A:

$\frac{1}{R_{eq}} = \frac{1}{R_1} + \frac{1}{R_2} + \frac{1}{R_3}$ ... so, in this circuit, $\frac{1}{R_{eq}} = \frac{1}{220} + \frac{1}{200} + \frac{1}{470}$

$$\frac{1}{R_{eq}} = \frac{1}{0.0117} \quad so \ R_{eq} = 85.67 \ ohms$$

Ohm's Law

Ohm's Law describes the relationship between voltage, current, and resistance, which are criteria used to characterize a given circuit. The difference in electrical potential (or voltage drop) between two different points in a circuit can be calculated by multiplying the current between the two points (I) and the total resistance of the electrical devices in the circuit between the two points (R).

ΔVoltage (V) = current $(I) \times$ resistance (R), where *V* is voltage (in volts), *I* is current (in amperes), and *R* is resistance (in ohms).

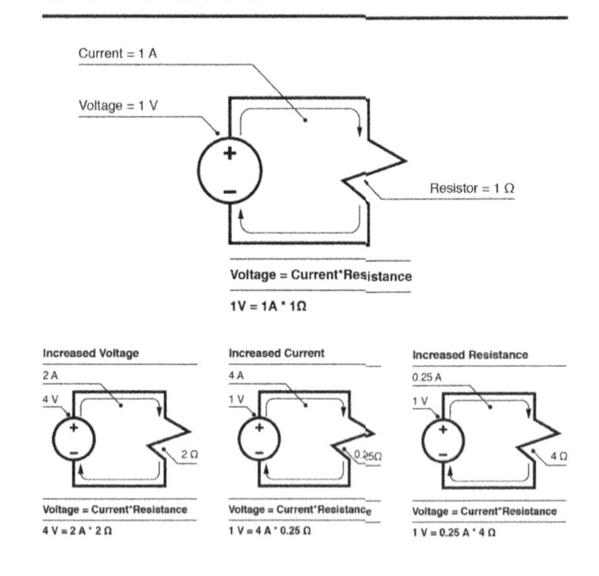

Alternatively, the following "Ohm's Triangle" is a useful tool to memorize the relationships governed by Ohm's Law. Test takers will need to memorize this equation for the exam. The triangle serves as a pictorial reminder and method to generate the correct relationships between voltage, current, and resistance.

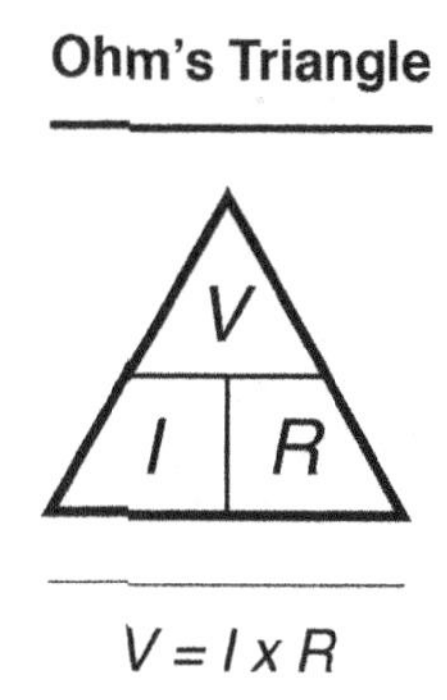

Recalling the standard Ohm's Law that $V = I \times R$ helps set up the basic triangle from which the other two equations can be visually transposed for those who find mathematically manipulating equations difficult.

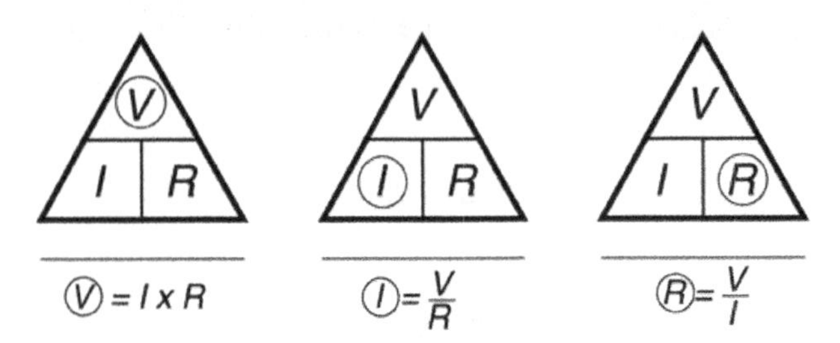

Example

Q. Which of the following equations is not an accurate representation of Ohm's Law?

a. $V = I \times R$

b. $R = \frac{V}{I}$

c. $I = \frac{V}{R}$

d. $V = \frac{R}{I}$

Explanation

Answer. D: Ohm's Law describes the relation between voltage and amperage, where voltage is a measure of electromotive potential much like potential energy of motion in kinetics. Amperage measures the electric current or the flow rate of 1 coulomb of electrons in a second. The basic law is $V = I \times R$, but this can be manipulated in the following ways:

$$I = \frac{V}{R} \quad \text{or} \quad V = I \times R \quad \text{or} \quad R = \frac{V}{I}$$

Series Circuits

In a **series circuit**, the electric charge passes consecutively through each device. When in series, charge passes through every light bulb. In a series circuit, there is a single pathway for electric current to flow, and all of the devices are added in succession to the same line. There are no branches coming off of this line, nor are there smaller loops within the circuit. The same electric current runs through each device, but the voltage drops as more devices or resistors are added to the string of connected devices.

While current doesn't change, voltage does drop after each resistor in such a way that the total voltage across the circuit is equal to the sum of the voltages across each device or resistor. An equivalent basic circuit or the equivalent total resistance in the circuit can be calculated by adding the specific resistance for each resistor together.

$$R_{equiv} = R_1 + R_2 + R_3 ...$$

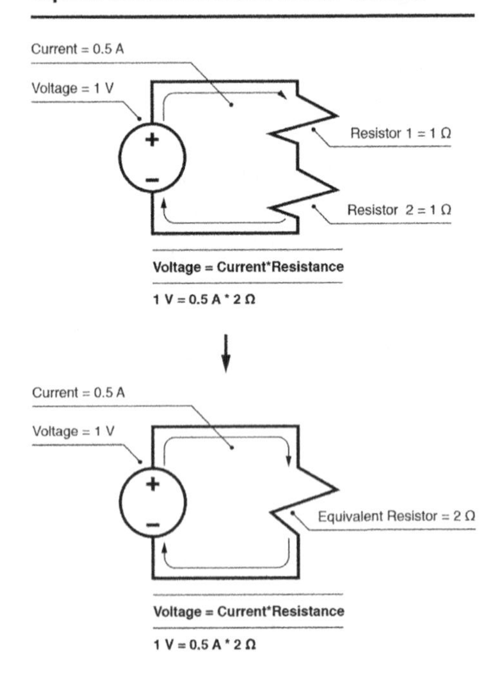

Current is the same at all locations along the circuit, whether measured at the battery, the first resistor, or after the last resistor. There is no accumulation or pileup of charge at any given location, and charge does not decrease or get used up by resistors such that there is any less charge at one point in the path compared to another.

Therefore:

$$I_{battery} = I_{resistor1} = I_{resistor2} = I_{resistor3} \cdots$$

Example

Q. Which of the following is true regarding series circuits?

a. The voltage drops across each resistor, but the current is the same in all of them.
b. The voltage is the same across each resistor, but the current drops after each of them.
c. The voltage drops across each resistor, and the current drops in each of them.
d. The voltage and current are the same across each of them in the series.

Explanation

Answer. B: When devices are connected in a series circuit, the voltage is the same across each device (or resistor) in the loop, but the current drops after each device. Recall that for lightbulbs connected in series, this is observed by a decrease in the brightness of each successive bulb. This is one of the disadvantages of a series circuit. In contrast, the current increases with the addition of each resistor in parallel circuits.

Parallel Circuits

When devices are connected in a **parallel circuit**, the charge passes through the external circuit and will only traverse one of the branches during its path back to the other terminal of the energy source. For example, if there are several light bulbs on separate branches connected in a parallel series, a single charge only passes through one of the light bulbs. Therefore, the voltage is the same across each resistor because each resistor is attached directly to the power source and the ground. In stark contrast to series circuits, as the number of devices or resistors increases in a parallel circuit, the overall current also increases. Somewhat counterintuitively, the addition of more resistors in a separate branch of the circuit actually decreases the overall resistance in the parallel circuit, so the current increases.

Here is an example image using an analogy of the flow rate on a tollway:

Like with series circuits, resistors in parallel can also be reduced to an equivalent circuit, but not by simply adding the resistances. The equivalent resistance (R_{eq}) is instead found by solving Ohm's Law for the current through each resistor, setting this value equal to the total current (I_t), and remembering that the voltages are all identical. Essentially, this yields an equation that shows that the inverse of the equivalent resistance of parallel resistors is equal to the sum of the inverses of the resistance of each leg of the parallel circuit.

Mathematically, that means:

$$I_t = \frac{V}{R_{eq}} = \frac{V}{R_1} + \frac{V}{R_2} \text{ or } \frac{1}{R_{eq}} = \frac{1}{R_1} + \frac{1}{R_2} \text{ so } R_{eq} = \frac{1}{\frac{1}{R_1} + \frac{1}{R_2}}$$

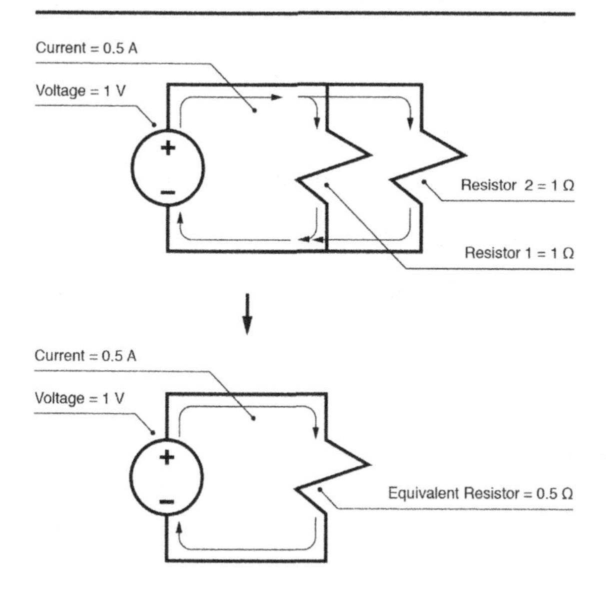

$$R_{equiv} = \frac{1}{\frac{1}{1\,\Omega} + \frac{1}{1\,\Omega}} = 0.5\,\Omega$$

As mentioned, the total resistance decreases as additional resistors are added to the parallel series, and the equivalent resistance is always lower than that of the smallest resistor.

Example

Q. Which of the following statements about parallel circuits is FALSE?

a. If the resistance is different in each load, more current passes through the load with the lower resistance.

b. If the resistance is different in each load, more current passes through the one with higher resistance.

c. If the resistance is the same in both loads, then the same amount of current passes through each one.

d. The voltage is the same across each resistor.

Explanation

Answer. B: In a parallel circuit, more charges will seek the path with least resistance. Because of this, more current passes through circuit paths with lower resistance. If the resistance is the same in both paths, then the same amount of current passes through each resistor.

Electrical Power

Electrical power in a circuit refers to the energy produced or absorbed, and it is expressed in watts (W). In a given circuit, certain components such as light bulbs consume electrical power and convert it to light/heat, while other components, for example, the battery, produce power. Power, in watts, is equal to the current from a voltage source in amperes multiplied by the voltage of that source:

$$\text{Power } (W) = \text{current } (I) \times \text{voltage } (V)$$

Using Ohm's Law and substitution, this relation can also be written in the following two ways:

$$W = I^2 R$$

$$W = \frac{V^2}{R}$$

Example

Q. Electric power can be measured in which of the following units?

a. Volts
b. Amperes
c. Ohms
d. Watts

Explanation

Answer. D: Watts are a measure of power or the rate at which electrical energy is used or converted into another type of energy. Watts are often converted to kilowatts or, when expressing power rates, kilowatt-hours.

AC Versus DC

The circuits previously described are **direct current (DC) circuits**, which are circuits wherein the voltage source has a constant value, and the current flows unidirectionally. This type of circuit is usually used for the wiring on ships, airplanes, and electronic devices. Batteries produce DC, and devices called rectifiers convert **alternating current (AC)** to DC. In contrast, voltage alternates over time in AC circuits, and current flow can switch directions. AC electricity is used in most land-based, heavy machinery and powers houses and buildings because large amounts of AC electricity can be transmitted over long distances with significantly less loss of power than with DC electricity.

Capacitors

Capacitors are devices that store electric charge and resist changes in voltage. They are made from two parallel plates made of conductive surfaces that are separated by a space or an insulating material such as ceramic or Teflon. Capacitors filter signals by blocking DC signals but permitting AC ones.

Capacitors

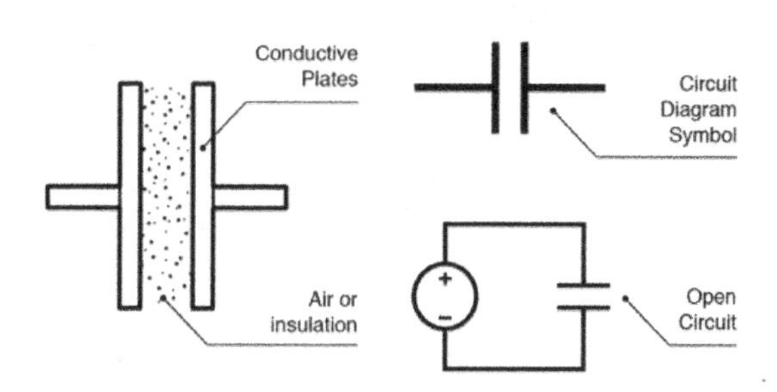

Example

Q. All EXCEPT which of the following statements are true regarding capacitors?

a. It is a set of parallel plates separated by a non-conducting material.

b. They don't stop ACs, but they do stop DCs.

c. The electric charges built up on either side of the capacitor resist change equal to the voltage of the power source.

d. The coils in the electric motors of AC equipment are examples of capacitors.

Explanation

Answer. D: The coils in the electric motors of AC equipment serve as inductors (not capacitors), which are typically coils of conducting wire in which a magnetic field is created by the electric current.

Examples of capacitors are ceramic, Teflon, and air because these are non-conducting materials.

Inductors

Inductors are coils of wire that develop a magnetic field in the presence of electrical flow. This magnetic field resists changes in the magnitude of changes in current in AC circuits. They can function as filters because they block AC signals but not DC, and they can separate signals of varying frequencies.

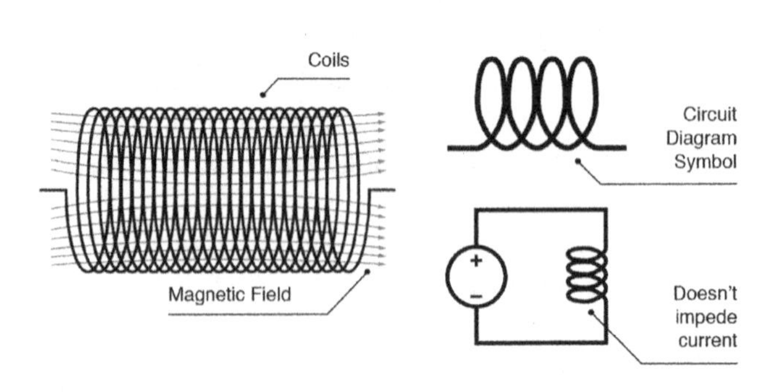

Electrical Diagrams

Test takers should be familiar with reading **block diagrams**, which depict electrical circuits with various symbols. The following figure includes frequently encountered symbols, and basic definitions for the important terms not yet covered are listed below the figure.

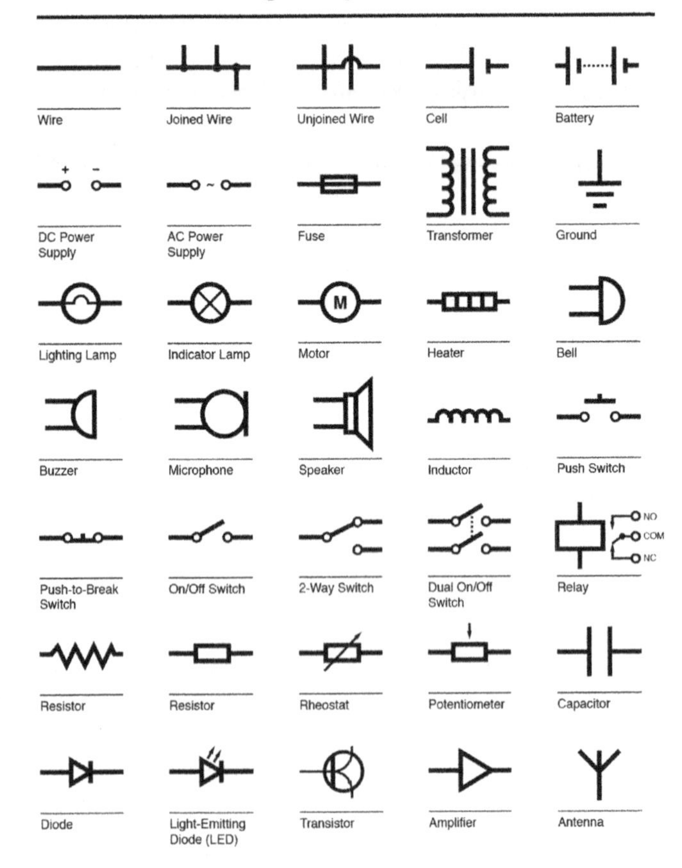

- **Cell**: The electronic unit that supplies current; batteries contain multiple cells.

- **Fuse**: A safety device that protects against current that surpasses a specified value by opening the circuit to prevent current flow.
- **Transformer**: A device that increases or decreases AC voltage by transferring energy via magnetic field induction between two coils of wire that are not electrically connected.
- **Ground**: The electrical connection to the earth, which enables the ground to serve as a zero-point reference for voltage against which other voltages in a circuit can be measured.
- **Transducer**: A device, such as a motor or lamp, that converts energy from one form into a different form.
- **Relay**: A type of switch that can open or close a circuit electromagnetically or electrically.
- **Rheostat**: A two-terminal variable resistor that controls current (such as lamp brightness).
- **Potentiometer**: A three-terminal variable resistor that controls voltage.
- **Diode**: The electrical equivalent of a valve; it permits electricity to flow in only one direction, indicated by the arrow in diagrams. Light-emitting diodes (LEDs) illuminate when current passes through.

- **Transistor**: A type of semiconductor (not quite a conductor yet not fully an insulator) that amplifies or switches electric signals or power.
- **Amplifier**: An electrical circuit contained in a device that increases the voltage, power, or current of a signal.

Example

Q. Which of the following best describes the function of and symbol for a diode?

a. It is a type of variable resistor that controls voltage, and it is represented as: 

b. It is a type of variable resistor that controls voltage, and it is represented as: 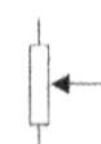

c. It ensures that current flows one way only, and it is represented as: 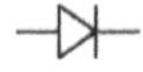

d. It ensures that current flows one way only, and it is represented as: 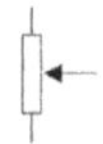

Explanation

Answer. C: Diodes function in electrical circuits much like valves do with fluid dynamics or in veins. They prevent current from flowing in the unintended direction and instead only permit flow in one direction.

The symbol is:

The arrow points in the direction of permitted current flow.

Auto and Shop Information

Types of Automotive Tools

There are several tools that are indispensable to the auto mechanic:

- **Wrenches**: The 9/16" and ½" fixed open wrench are among the most common and necessary tools for the vehicle mechanic. A complete ratchet set is also a necessary kit.
- **Vice grips**: Medium-sized vice grips lock into place in order to hold onto something, which enables the mechanic to grip items with a grip-locking feature.
- **Channel lock pliers**: The jaws on these pliers are positioned at a horizontal right-angle, which enables the mechanic to grab onto a nut or bolt that would be hard to turn.
- **Flat head screwdrivers**: One large screwdriver, preferably about 10 inches long, is an important tool that can do double duty as a small pry bar. A smaller flat head screwdriver is good for the more diminutive slotted screws. The 5-inch screwdriver is also useful for prying things up when a smaller tool is required.

- **Phillips-head screwdrivers**: the four slotted "Phillips" screws are encountered throughout a vehicle and various sized Phillips-head screwdrivers are important for the mechanic to have on hand to effectuate repairs.

Example

Q. What are channel lock pliers?

a. A fixed set of pliers that can tune the channel of a car's radio

b. A funnel that channels oil into the engine of a car

c. A type of adjustable pliers that have a right-angle positioned jaw

d. A straight jaw pliers that are used to unlock car doors

Explanation

Answer. C: The jaws on these pliers are positioned at a horizontal right-angle, which enables the mechanic to grab onto a nut or bolt that would be hard to turn. The word *channel* in channel lock pliers denotes the tongue and groove adjustment feature on the pliers.

Automotive Systems

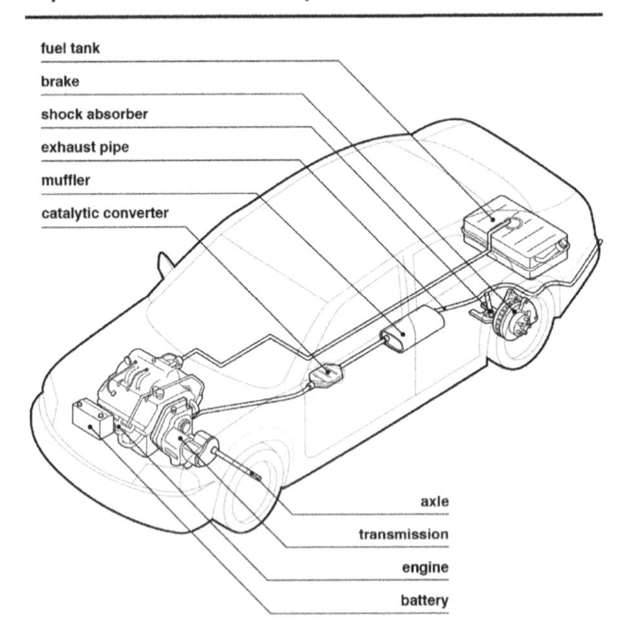

The main source of power for vehicles around the world is still the gasoline combustion engine. Primary systems of the vehicle must be maintained and repaired. The primary systems of a vehicle include, but are not limited to:

The Engine System

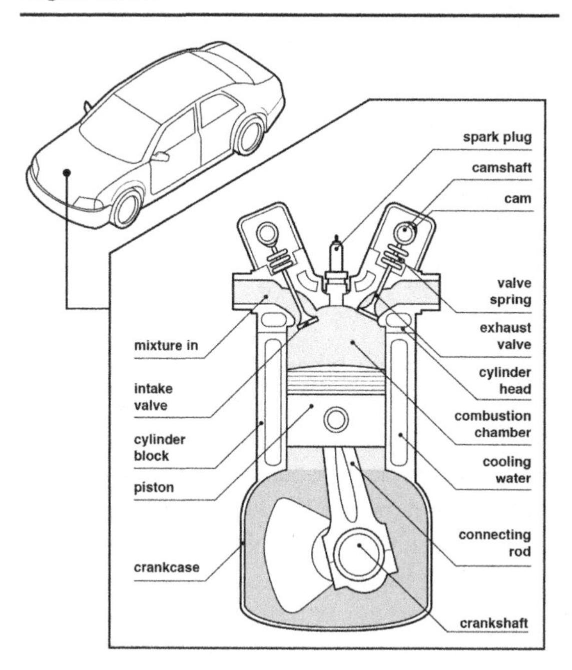

Primary engine systems for a gasoline combustion-based motor include:

- Engine block

- Pistons
- Spark plugs
- Rocker arms
- Cams
- Valves
- Fuel injectors
- Carburetors
- Electrical wires
- Starters
- Gaskets

The Ignition System

The car's **ignition key**, when turned in traditional cars or pushed in some newer vehicles, completes an electrical connection from the car's battery and sends current to the ignition module that controls the electrical ignition motor. The electrical **ignition motor** turns the crankshaft, which then initiates the up and down motion of the cylinders while the valves above the cylinders inject gas into the piston chambers. The **spark plugs** ignite the gas and create combustion in the cylinders. **Combustion** inside the engine cylinders is a self-perpetuating power system, as long as gas continually feeds into the cylinders. The **alternator** is an electrical generator powered by the gas engine.

The Electrical System

The **battery** is the heart of the vehicle's electrical system; it is where the electrical system begins and ends. Another essential part of the electrical system is a unit known as the **alternator**. The alternator is driven by the crankshaft. The spinning of the alternator by the crankshaft produces electricity for the vehicle while the engine is running. If the alternator fails, then it must be replaced. A **tensioner** device is used to keep tension on the pulley bands. The rubber bands use a pulley system, so it is important for the mechanic to use a pry bar to get tension on the rubber bands that go over the pulleys.

The Drive System

While the engine is running, the flywheel turns the crankshaft. This moves the pistons in their respective cylinders. Any components located in the lower end the engine near the block, or components located inside the engine or located just below the engine block, are difficult to repair and require many tools. The use of an auto shop with a vehicle lift and full array of specialized tools is required to effectuate repairs to the lower part of the engine. The car's transmission transmits power produced by the engine to the wheels. The transmission is a complex component that requires special training to carry out needed repairs.

The Lubrication System

The oil contained in an oil pan below the engine is the "life blood" vat of the vehicle's lubrication system. When the pistons move, and the combustion engine starts operating, it starts the mechanical operation of the oil pump, which then circulates oil from the pan located below the engine. The oil moves up to the valves above. The oil coats the pistons then drops back into the oil pan. Lubrication prevents heat damage to the engine due to friction. The oil pump sits inside the oil pan. To replace the oil pump, the mechanic must drop the oil pan down by using different sized wrenches, and then the new oil pump can be reconnected.

Cooling System

In order to cool the engine, water circulates in channels throughout the engine. As the engine runs, the engine fan takes in air through the radiator. As the air passes by the fins of the radiator, the flowing water cooled by the air cools the engine. Cooled water flows through the engine jacket.

The Exhaust System

The **exhaust manifold**, also known as a **front pipe**, is designed to take gases and gas byproducts from the engine cylinders and channel them into the exhaust. Then the exhaust gas travels down an exhaust pipe to a catalytic converter, which is a filter designed to reduce harmful pollutant gases. The center section, also known as a **muffler** or **silencer**, is designed to reduce noise. The tailpipe further reduces noise and funnels exhaust gases away from the vehicle.

The Brake System

When the brake is depressed, the brake fluid flows through the master cylinder and then to each wheel's break cylinder. The resulting fluid pressure causes physical pressure, which creates friction. A hydraulic brake system is an extension of Pascal's Law, which states: "Pressure exerted anywhere in a contained incompressible fluid is distributed equally in all direction throughout the fluid."

Suspension System

The **suspension system** of a vehicle helps provide a smooth ride when driving. Vibration can cause a host of maladies within a car and can literally shake a car to pieces if uncontrolled. The less vibration that a vehicle experiences, the fewer the necessary repairs. The other important factors associated with the suspension system is that the system keeps the vehicle upright and supports the vehicle from excessively tilting and swaying when the vehicle is turning, accelerating, and braking.

The Steering System
All vehicles have a **steering wheel** that turns the front wheels of the vehicle. The front end of a vehicle usually has a lot of weight because the engine is located in front. The steering mechanisms are usually powered by an **engine-driven pump** and a **hydraulic system**. With an older rack and pinion steering system, the rotational movement of the steering wheel is transferred to the frame mounted rack and pinion steering assembly. The **rack** and **pinion mechanism** offer precise steering abilities but require more effort to move the wheels. The main steering components consist of a steering wheel, steering shaft, a steering rack or steering box, and arm mechanism.

Example
Q. A master cylinder is an important component in which vehicular system?

a. Electrical
b. Transmission
c. Engine
d. Brake

Explanation
Answer. D: The brake system is arguably the most important system in any vehicle. If a car cannot "go," it is usually not an immediate safety issue. However, if the brakes fail to stop a vehicle's motion, then

physical damage and serious injury or death can result. If the brake fluid is not present in the master cylinder, then the brakes will not operate, as brakes depend on the principles of fluid dynamics to work.

Automotive Components

There are several essential components of a vehicle that require replacement including:

- Brake Pads
- Drum brakes
- Valve cover gaskets
- Tires
- Lights
- Hoses
- Wires
- Electrical connectors

Example

Q. Which vehicle component often needs to be replaced on a regular basis?

a. Rack and pinions
b. Brake pads
c. Emergency brakes
d. Window seals

Explanation

Answer. B: Brake pads often wear out from normal driving and must be replaced. Over time, the brake pads will totally wear down, causing the brake caliper to contact the brake disc, which leads to brake disc damage.

Procedures for Automotive Troubleshooting and Repair

The vehicle operator and vehicle mechanic must be cognizant of any puddles of liquid that have formed under the vehicle. Leaking brake fluid necessitates immediate repair. **Antifreeze** is greenish-yellow and has a distinct smell. If an antifreeze leak is suspected, the coolant vat should be checked and if low, there may be a leak. A leaking **oil pan** is usually detected when a car moves from a parking spot and there is a significant spot of oil where the car was parked. Any amount of oil detected on the pavement can denote a dangerous and costly condition that must be repaired. **Water condensation** from the vehicle's exhaust tailpipes is not a concern. Most vehicles have water drips from the tailpipes when the vehicle is first turned on and operated for the first few minutes.

Example
Q. If a puddle of liquid is observed under a vehicle, what is the correct action?

a. Mark the day on the calendar when the car should be serviced.
b. Identify the liquid under the car and act accordingly.
c. Taste the liquid. If the liquid tastes like oil, the oil pan needs replacing.
d. Water from the air conditioner drains the coolant, indicating the radiator needs filling.

Explanation
Answer. B: Identifying the type of liquid under a car can be the difference between driving away with no consequence or burning out the engine and causing an expensive repair. If the puddle is oil, then the engine may be in danger of burning out. If the puddle is just water from condensation produced by the vehicle's air conditioning system, then no action needs be taken because this water is normal.

The Vehicle Lift

Proper operational knowledge of a car "lift" function is essential for the vehicle mechanic. A **car lift** enables the mechanic to operate under the car without having to lay supine underneath the vehicle. Moreover, the lift permits the removal and installation of larger components under a vehicle. Working under a four-thousand-pound car can be extremely dangerous if the mechanic is not familiar with the safe operation of the car lift.

An adequately equipped machine shop or vehicle repair shop contains many tools including the following:

- Metal Inert Gas
- Sawzall™
- Stout hand drill
- Cobalt wrenches
- Wire strippers
- Rubber mallet
- Assorted steel nuts, bolts, and washers
- Carbide blade chop saw
- Scrap bin
- Multi-drawer toolbox
- Metal files
- Simple shelving
- Dial calipers
- Angle grinder
- Filtration system
- Tungsten Inert Gas (TIG)
- Mill drills
- Lathe
- Air compressor

Example

Q. If a mechanic must cut off a car's exhaust tailpipe, what tool should the mechanic choose?

a. A lathe
b. A chop saw
c. A Sawzall™
d. An axe

Explanation

Answer. C: The battery powered Sawzall™ is a tool that can cut metal easily and also get into tight spaces. It is hand-operated, and it gives the mechanic the flexibility to reach into tight spaces in the vehicle.

Construction Procedures

Concrete is an advancement that has revolutionized the ability to build comfortable residential structures using materials that are fantastically resilient to the elements. The structures built with brick, stone, and concrete can outlive the length of a human life. There are brick structures from 500 years ago that are still standing today, and some stone and concrete built houses can last for thousands of years.

Example

Q. When building a roof for a house, which procedure is the correct construction practice?

a. Lay tar paper over the bare plywood

b. Create space between the shingles so water flows into the side ducts

c. Place bricks under each shingle for extra support

d. Use short nails through the shingles to prevent damage

Explanation

Answer. A: Laying tar paper over the plywood on a roof protects the plywood, which would otherwise absorb water. The tar paper is essentially another layer of waterproofing under the shingles.

Constructing Brick Walls

To **construct a brick wall**, the mortar is spread completely over the top of each brick with a trowel or spatula tool. These flat tools are used to smooth out the mortar used between the bricks and to wipe away excess mortar that spills over the sides of the brick. The brick wall should be free of mortar on the sides of the bricks. One layer of bricks should be laid down and each brick should be tapped with the end of the trowel's handle. A "spoon" tool is used to create a concave indentation in the concrete that surrounds each brick. A bubble leveler is a very important tool and should be used consistently through the building of a brick wall.

Example

Q. Which of the following is used in brick laying?

a. Cement
b. Trommel
c. Plaster
d. Trowel

Explanation

Answer. D: A trowel is used to smooth and wipe away extra mortar.

The Roof

A waterproof, weatherproof **roof** on top of a structure is an essential part of the structure that must be resilient. There are just a few important tools and techniques to employ for the construction of a new roof. A hammer, a bucket of nails, a sufficient number of shingles, and tar paper are the main materials that are required. A flat shovel or the front of the spade can be used to get underneath the old shingles of the roof to shovel them off and expose the base roof plywood. Then tar paper is put over the plywood using a staple gun and then shingles are added from the bottom of the roof to the top. Roofing nails with big heads and shafts long enough to go through the tar paper, shingles, and plywood must be used.

Example

Q. How are shingles added to a roof?

a. From left to right
b. From right to left
c. From bottom to top
d. From top to bottom

Explanation

Answer. C: Shingles should be added to the roof from the bottom to the top. This ensures that there are no gaps for water to trickle through as runs down the roof.

Foundation

A poured concrete **foundation** is standard for the first stage in building a structure in locations where temperatures can often go below freezing. **Rebar**—iron/metal rods that extend through the concrete—are used as a reinforcement for the foundation. **Wood frames** can serve as molds for cement walls and are easily built using wooden beams, a circular saw, a hammer, and nails. Cement does not stick to wood, so wood is a good foundation framing material. After the concrete hardens in the frame, then the wood easily comes off away from the concrete. Concrete wall thickness is determined by the size and design of the intended structure but 18" thickness is the norm. The foundation should be at least 46 inches below the ground line.

Example

Q. Which of the following is used as reinforcement in concrete?

a. Lag bolts
b. Rebar
c. 2" × 4" boards
d. Galvanized pipe

Explanation

Answer. B: Rebar is a long metal rod. These rods are used in foundations to provide reinforcement to the concrete.

Portals: Door Frames and Window Frames

Typically, either 2" x 4" or 2" x 6" boards form a windowsill. Four boards support the sill; these boards are also referred to as **studs**. Trimmer studs extend vertically from the bottom horizontal stud and these form the sides of the window frame and are nailed into the king stud, which is one of the studs that make up the main frame of the wall. The trimmer studs support the **header assembly**, which is the top of window frame. The header assembly is four different individual pieces of wood and often must conform to building code regulations within the structure's municipality.

A **door frame** is very similar except that there is no triple support from below because the door goes all the way down to the floor. The trimmer studs on either side of the portal make up the door frame sides. The trimmer studs support the top or header of door. The door header should have rows of three nails every sixteen inches. The header of a door should be strong because it needs to support anything that is above it.

Example

Q. Which important part of a window frame must conform to building code?

a. The sill
b. The trimmer
c. The stud
d. The header

Explanation

Answer. D: The top of the window frame, the header, often must bear weight and is under the municipal umbrella of construction rules. This top part of the window frame is also the most complex, consisting of the most wooden boards or other material used.

Mechanical Comprehension

Mechanical Comprehension Section

The test problems in the Mechanical Comprehension section of the exam focus on understanding physical principles, but they are qualitative in nature rather than quantitative. This means the problems involve predicting the behavior of a system (such as the direction it moves) rather than calculating a specific measurement (such as its velocity).

Example

Mechanical Comprehension Sample Test Problem

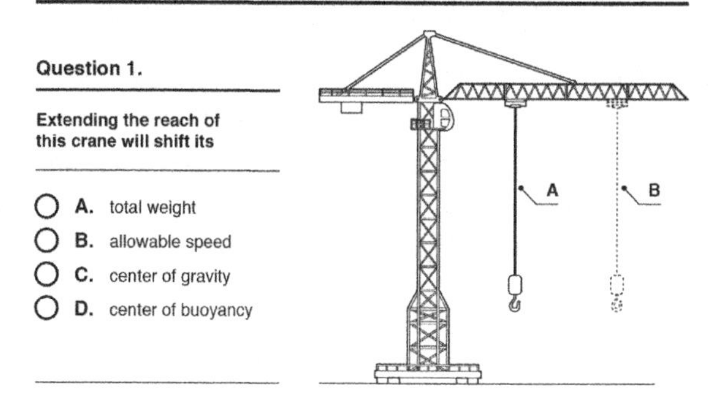

Explanation

The correct answer is *C, center of gravity*. In this sample problem, it's easy to guess the correct answer simply by eliminating the rest. Choice *A* is incorrect because moving the load out along the crane's boom won't change its weight, just like moving a bodybuilder's arm that's holding a dumbbell won't change the combined weight of the bodybuilder and the dumbbell. Choice *B* is incorrect because the crane isn't moving. That leaves Choices *C* and *D*, but *D* is incorrect because buoyancy is only involved in systems with a liquid (the buoyancy of air is negligible). Therefore, through the process of elimination, *C* is the correct answer.

Review of Physics and Mechanical Principles

The matter in the universe (atoms and molecules) is characterized in terms of its **mass**, which is measured in kilograms in the **International System of Units** (**SI**). The amount of mass that occupies a given volume of space is termed **density**.

Mass occupies space, but it's also a component that inversely relates to acceleration when a force is applied to it. This **force** is the application of energy to an object with the intent of changing its position (mainly its acceleration).

To understand **acceleration**, it's necessary to relate it to displacement and velocity. The **displacement** of an object is simply the distance it travels. The **velocity** of an object is the distance it travels in a unit of time, such as miles per hour or meters per second:

$$Velocity = \frac{Distance\ Traveled}{Time\ Required}$$

There's often confusion between the words "speed" and "velocity." Velocity includes speed *and* direction. For example, a car traveling east and another traveling west can have the same speed of 30 miles per hour (mph), but their velocities are different. If movement eastward is considered positive, then movement westward is negative. Thus, the eastbound car has a velocity of 30 mph while the westbound car has a velocity of -30 mph.

The fact that velocity has a **magnitude** (speed) and a direction makes it a vector quantity. A **vector** is an arrow pointing in the direction of motion, with its length proportional to its magnitude.

Vectors can be added geometrically as shown below. In this example, a boat is traveling east at 4 knots (nautical miles per hour) and there's a current of 3 knots (thus a slow boat and a very fast current). If the boat travels in the same direction as the current, it gets a "lift" from the current and its speed is 7 knots. If the boat heads *into* the current, it has a forward speed of only 1 knot (4 knots – 3 knots = 1 knot) and makes very little headway. As shown in the figure below, the current is flowing north across the boat's path. Thus, for every 4 miles of progress the boat makes eastward, it drifts 3 miles to the north.

Working with Velocity Vectors

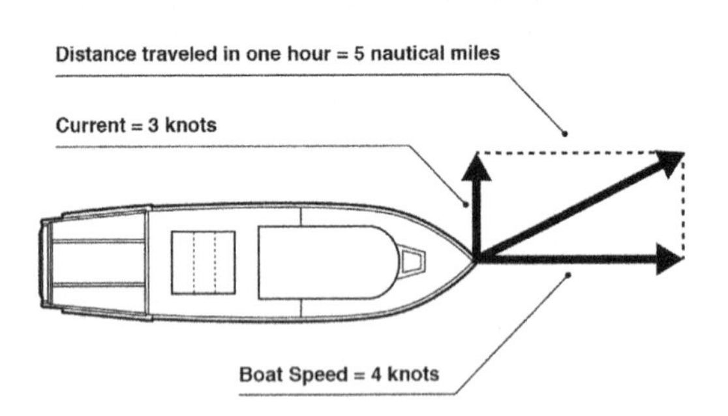

The total distance traveled is calculated using the *Pythagorean Theorem* for a right triangle, which should be memorized as follows:

$$a^2 + b^2 = c^2 \text{ or } c = \sqrt{a^2 + b^2}$$

Of course, the problem above was set up using a Pythagorean triple (3, 4, 5), which made the calculation easy.

The **acceleration** of an object is the change in its velocity in a given period of time:

$$Acceleration = \frac{Change\ in\ Velocity}{Time\ Required}$$

Example

Q. A car is traveling at a constant velocity of 25 m/s. How long does it take the car to travel 45 kilometers in a straight line?

a. 1 hour
b. 3600 seconds
c. 1800 seconds
d. 900 seconds

Explanation

Answer. C: The answer is 1800 seconds:

$$Desired\ distance\ in\ km \times conversion\ factor\ (m\ to\ km)) / current\ velocity\ in\ \frac{m}{s}$$

$$\left(45\ km \times \frac{1000\ m}{km}\right) \Big/ 25\frac{m}{s} = 1800\ seconds$$

Newton's Laws

Isaac Newton's three laws of motion describe how the acceleration of an object is related to its mass and the forces acting on it. The three laws are:

- Unless acted on by a force, a body at rest tends to remain at rest; a body in motion tends to remain in motion with a constant velocity and direction.
- A force that acts on a body accelerates it in the direction of the force. The larger the force, the greater the acceleration; the larger the mass, the greater its inertia (resistance to movement and acceleration).
- Every force acting on a body is resisted by an equal and opposite force.

To understand Newton's laws, it's necessary to understand forces. These forces can push or pull on a mass, and they have a magnitude and a direction. Forces are represented by a vector, which is the arrow lined up along the direction of the force with its tip at the point of application. The magnitude of the force is represented by the length of the vector.

The figure below shows a mass acted on or "pushed" by two equal forces (shown here by vectors of the same length). Both vectors "push" along the same line through the center of the mass, but in opposite directions. What happens?

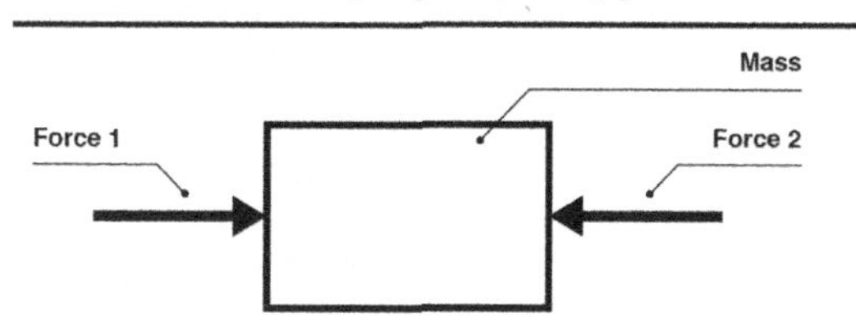

According to Newton's third law, every force on a body is resisted by an equal and opposite force. In the figure above, Force 1 acts on the left side of the mass. The mass pushes back. Force 2 acts on the right side, and the mass pushes back against this force too. The net force on the mass is zero, so according to Newton's first law, there's no change in the **momentum** (the mass times its velocity) of the mass. Therefore, if the mass is at rest before the forces are applied, it remains at rest. If the mass is in motion with a constant velocity, its momentum doesn't change. So, what happens when the net force on the mass isn't zero, as shown in the figure below?

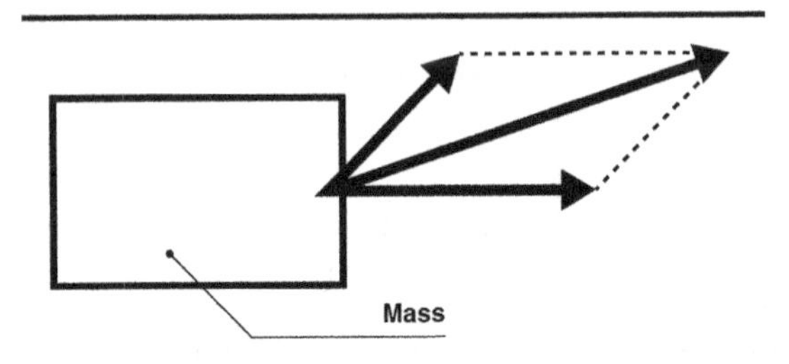

Notice that the forces are vector quantities and are added geometrically the same way that velocity vectors are manipulated.

Here in the figure above, the mass is pulled by two forces acting to the right, so the mass accelerates in the direction of the net force. This is described by Newton's second law:

Force = Mass x Acceleration

The force (measured in *newtons*) is equal to the product of the mass (measured in kilograms) and its acceleration (measured in meters per second squared or meters per second, per second). A better way to look at the equation is dividing through by the mass:

Acceleration = Force/Mass

This form of the equation makes it easier to see that the acceleration of an object varies directly with the net force applied and inversely with the mass. Thus, as the mass increases, the acceleration is reduced for a given force. To better understand, think of how a baseball accelerates when hit by a bat. Now imagine hitting a cannonball with the same bat and the same force. The cannonball is more massive than the baseball, so it won't accelerate very much when hit by the bat.

In addition to forces acting on a body by touching it, gravity acts as a force at a distance and causes all bodies in the universe to attract each other. The **force of gravity** (F_g) is proportional to the masses of the two objects (m and M) and inversely proportional to the square of the distance (r^2) between them (and G is the proportionality constant). This is shown in the following equation:

$$F_g = G\frac{mM}{r^2}$$

Another way to understand Newton's second law is to think of it as an object's change in momentum, which is defined as the product of the object's mass and its velocity:

Momentum = Mass x Velocity

Example

Q. A spaceship with a mass of 100,000 kilograms is far away from any planet. To accelerate the craft at the rate of 0.5 m/sec^2, what is the rocket thrust?

a. 98.1 N
b. 25,000 N
c. 50,000 N
d. 75,000 N

Explanation

Answer. C: The answer is 50,000 N. The equation $F = ma$ should be memorized. All of the values are given in the correct units (kilogram-meter-second) so plug them in to the equation:

$$F = 100{,}000 \text{ kg} \times 0.5 \frac{\text{m}}{\text{s}^2} = 50{,}000 \text{ N}$$

Projectile Motion

What happens when a bullet is fired from the top of a hill using a rifle held perfectly horizontal? Ignoring air resistance, its horizontal velocity remains constant at its muzzle velocity. Its vertical velocity (which is zero when it leaves the gun barrel) increases because of gravity's acceleration. Each passing second, the bullet traces out the same distance horizontally while increasing distance vertically (shown in the figure below). In the end, the projectile traces out a **parabolic curve**.

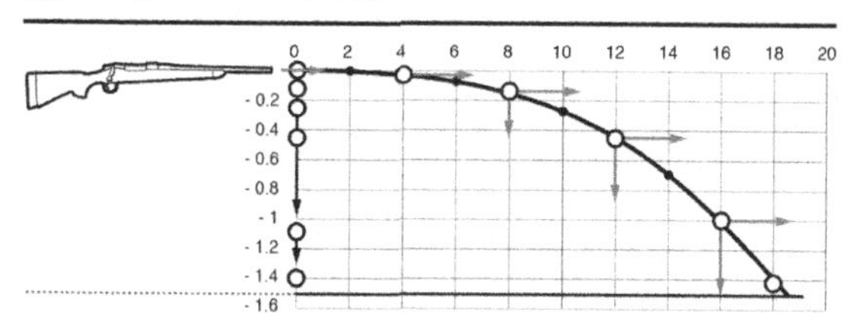

The horizontal velocity is no longer constant because the initial velocity of the projectile is continually reduced by the resistance of the air. This is a complex problem in fluid mechanics, but it's sufficient to note that that the projectile doesn't fly as far before landing as predicted from the simple theory.

The vertical velocity is also reduced by air resistance. However, unlike the horizontal motion where the propelling force is zero after the cannonball is fired, the downward force of gravity acts continuously. The downward velocity increases every second due to the acceleration of gravity. As the velocity increases, the resisting force (called **drag**) increases with the square of the velocity. If the projectile is fired or dropped from a sufficient height, it reaches a terminal velocity such that the upward drag force equals the downward force of gravity. When that occurs, the projectile falls at a constant rate.

Example

Q. A football is kicked so that it leaves the punter's toe at a horizontal angle of 45 degrees. Ignoring any spin or tumbling, at what point is the upward vertical velocity of the football at a maximum?

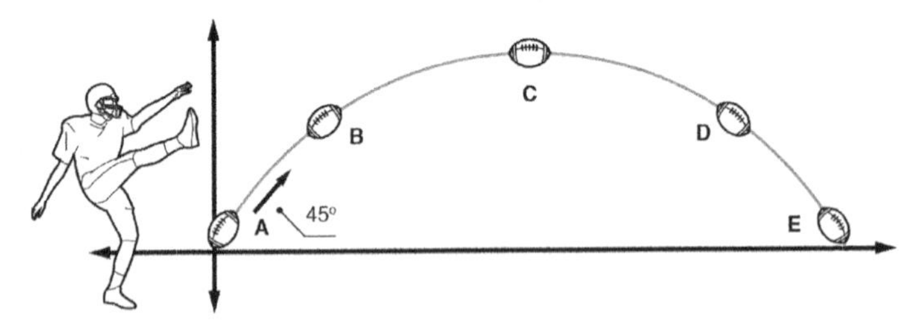

a. At Point A
b. At Point C
c. At Points B and D
d. At Points A and E

Explanation

Answer. A: The answer is that the upward velocity is at a maximum when it leaves the punter's toe. The acceleration due to gravity reduces the upward velocity every moment thereafter. The speed is the same at points A and E, but the velocity is different. At point E, the velocity has a maximum *negative* value.

Angular Momentum

In the previous examples, all forces acted through the center of the mass, but what happens if the forces aren't applied through the same line of action, like in the figure below?

A Mass Acted on by Forces Out of Line with Each Other

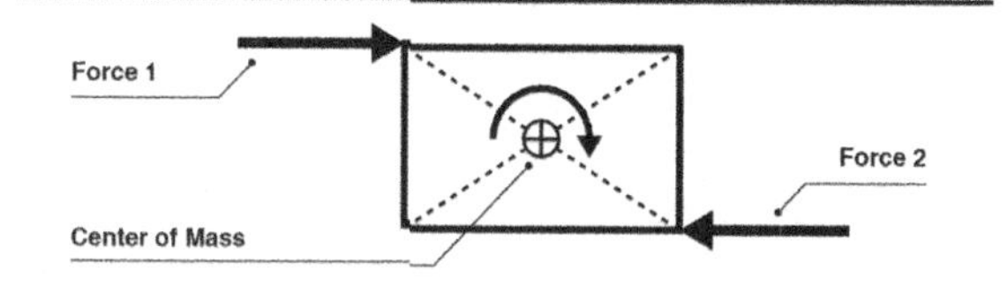

When this happens, the two forces create **torque** and the mass rotates around its center of gravity. In the figure above, the center of gravity is the center of the rectangle (**Center of Mass**), which is determined by the two, intersecting main diagonals. The center of an irregularly shaped object is found by hanging it from two different edges, and the center of gravity is at the intersection of the two "plumb lines."

Example

Q. The skater is shown spinning in Figure (a), then bringing in her arms in Figure (b). Which sequence accurately describes what happens to her angular velocity?

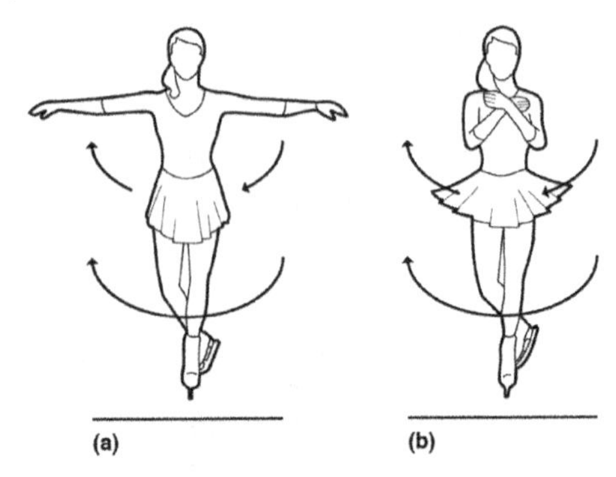

a. Her angular velocity decreases from (a) to (b)
b. Her angular velocity doesn't change from (a) to (b)
c. Her angular velocity increases from (a) to (b)
d. It's not possible to determine what happens to her angular velocity if her weight is unknown.

Explanation

Answer. C: The answer is her angular velocity increases from (a) to (b) as she pulls her arms in close to her body and reduces her moment of inertia.

Potential and Kinetic Energy

The **potential energy** of an object is equal to the work that's required to lift it from its original elevation to its current elevation. This is calculated as the weight of the object or its downward force (mass times the acceleration of gravity) multiplied by the distance (*y*) it is lifted above the reference elevation or "datum." This is written:

$$PE = mgy$$

The mechanical or **kinetic energy** of a system is related to its mass and velocity:

$$KE = \frac{1}{2}mv^2$$

The **total energy** is the sum of the kinetic energy and the potential energy, both of which are measured in foot-pounds or newton meters.

The conversion between potential and kinetic energy works the same way for a pendulum. If it's raised and held at its highest position, it has maximum potential energy but zero kinetic energy.

Potential and Kinetic Energy for a Swinging Pendulum

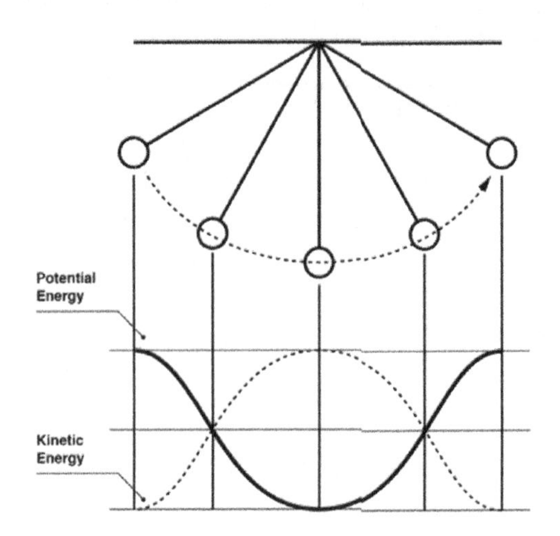

Work

The released potential energy of a system can be used to do **work**.

For instance, most of the energy lost by letting a weight fall freely can be recovered by hooking it up to a pulley to do work by pulling another weight back up (as shown in the figure below).

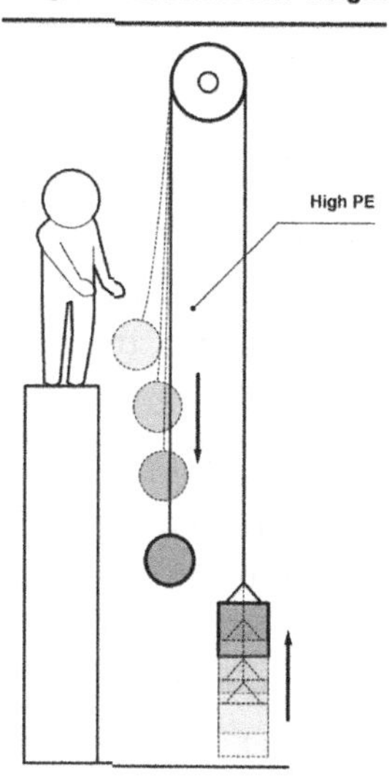

In other words, the potential energy expended to lower the weight is used to do the work of lifting another object. Of course, in a real system, there are losses due to friction. The action of pulleys will be discussed later in this study guide.

Since **energy** is defined as *the capacity to do work*, energy and work are measured in the same units:

$$Energy = Work = Force \times Distance$$

Force is measured in **newtons** (*N*). Distance is measured in meters. The units of work are **newton meters** *(N·m)*. The same is true for kinetic energy and potential energy.

Another way to store energy is to compress a spring. Energy is stored in the spring by stretching or compressing it. The work required to shorten or lengthen the spring is given by the equation:

$$F = k \times d$$

Here, "d" is the length in meters and "k" is the resistance of the spring constant (measured in N·m), which is a constant as long as the spring isn't stretched past its elastic limit. The resistance of the spring is constant, but the force needed to compress the spring increases with each millimeter it's pushed.

The potential energy stored in the spring is equal to the work done to compress it, which is the total force times the change in length. Since the resisting force of the spring increases as its displacement increases, the average force must be used in the calculation:

$$W = PE = F \times d = \frac{1}{2}(F_i + F_f)d \times d$$
$$= \frac{1}{2}(0 + F_f)d \times d = \frac{1}{2}Fd^2$$

The potential energy in the spring is stored by locking it into place, and the work energy used to compress it is recovered when the spring is unlocked. It's the same when dropping a weight from a height—the energy doesn't have to be wasted. In the case of the spring, the energy is used to propel an object.

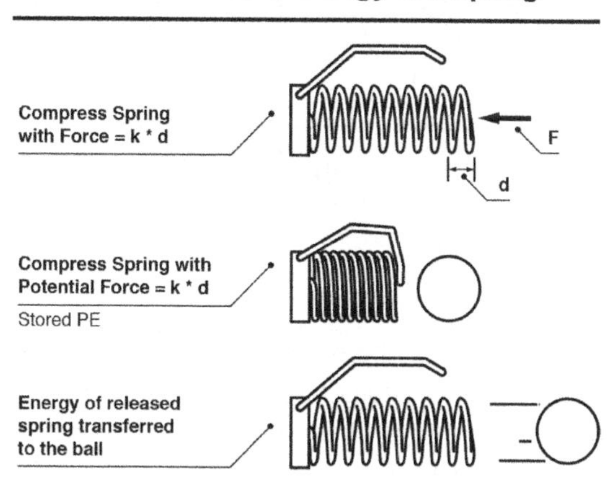

Pushing a block horizontally along a rough surface requires work. In this example, the work needs to overcome the force of friction, which opposes the direction of the motion and equals the weight of the block times a **friction factor** (*f*). The friction factor is greater for rough surfaces than smooth surfaces, and it's usually greater *before* the motion starts than after it has begun to slide. These terms are illustrated in the figure below.

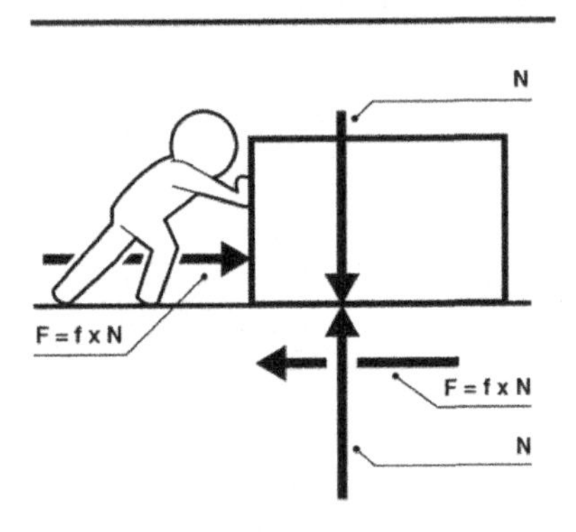

Heat energy can also be created by burning organic fuels, such as wood, coal, natural gas, and petroleum. All of these are derived from plant matter that's created using solar energy and photosynthesis. The chemical energy or *"heat"* liberated by the combustion of these fuels is used to warm buildings during the winter or even melt metal in a foundry. The heat is also used to generate steam, which can drive engines or turn turbines to generate electric energy.

The amount of work needed to raise the temperature of water was measured by an English physicist (and brewer) named James Prescott Joule. The way that Joule measured the mechanical equivalent of heat is illustrated in the figure below. Through a long series of repeated measurements, Joule showed that 4186 N·m of work was necessary to raise the temperature of one kilogram of water by one degree Celsius, no matter how the work was delivered.

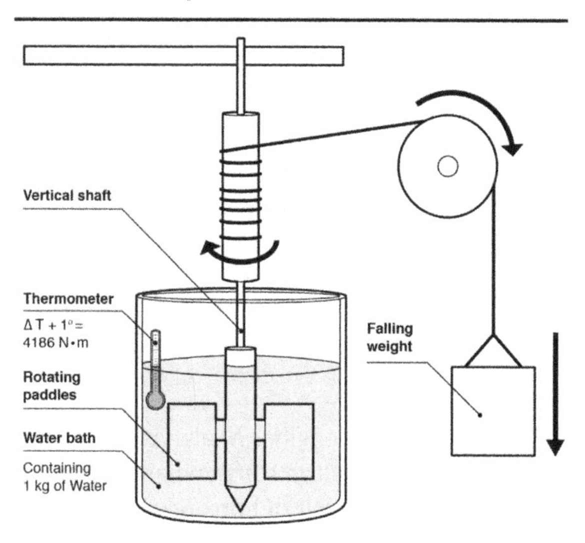

In recognition of this experiment, the newton meter is also called a **joule**.

Power

Power is defined as the rate at which work is done, or the time it takes to do a given amount of work. In the International System of Units (SI), work is measured in **newton meters** (*N·m*) or **joules** (*J*). Power is measured in joules/second or *watts (W)*.

To provide a measurement that the miners would be familiar with, Watt and Boulton referenced the power of their engines with the "power of a horse."

Watt's measurements showed that, on average, a well-harnessed horse could lift a 330-pound weight 100 feet up a well in one minute (330 pounds is the weight of a 40-gallon barrel filled to the brim). Remembering that power is expressed in terms of energy or work per unit time, horsepower came to be measured as:

$$1\,HP = \frac{100\,feet \times 330\,pounds}{1\,minute} \times \frac{1\,minute}{60\,seconds} = 550\,foot\,pounds/second$$

Hundreds of millions of engines of all types have been built since Watt and Boulton started manufacturing their products, and the unit of horsepower has been used throughout the world to this day. Of course, modern technicians and engineers still need to convert horsepower to watts to work with SI units. An approximate conversion is *1 HP = 746 W*.

A question that's often asked is, "How much energy is expended by running an engine for a fixed amount of time?" This is important to know when planning how much fuel is needed to run an engine. For example, how much energy is expended in running the new Cadillac at maximum power for 30 minutes?

In this case, the energy expenditure is approximately 240 kilowatt hours. This must be converted to joules, using the conversion factor that one watt equals one joule per second:

$$240{,}000\ W\ hours \times \frac{3600\ seconds}{1\ hour} = 8.64(10)^8\ joules$$

So how much gasoline is burned? Industrial tests show that a gallon of gasoline is rated to contain about 1.3 x 10^8 joules of energy. That's 130 million joules per gallon. The gallons of gasoline are obtained by dividing:

$$\frac{8.64(10)^8 J}{1.3(10)^8 J/gallon} = 6.65\ gallons \times \frac{3.8\ liters}{gallon} = 25.3\ liters$$

The calculation has now come full circle. It began with power. Power equals energy divided by time. Power multiplied by time equals the energy needed to run the machine, which came from burning fuel.

Example

Q. A gas with a volume V_1 is held down by a piston with a force of F newtons. The piston has an area of A. After heating the gas, it expands against the weight to a volume V_2. What was the work done?

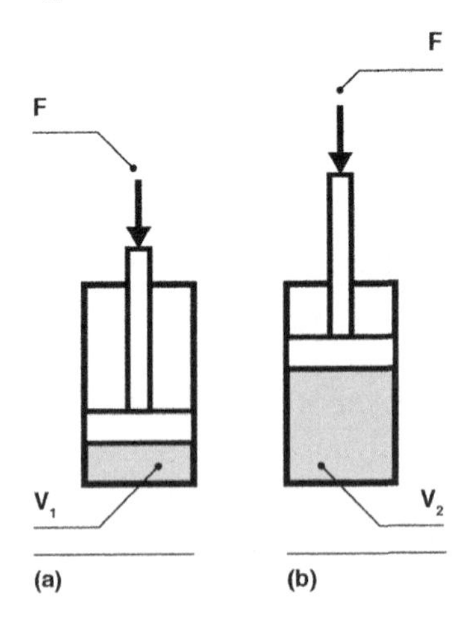

a. $\frac{F}{A}$

b. $\frac{F}{A} \times V_1$

c. $\frac{F}{A} \times V_2$

d. $\frac{F}{A} \times (V_2 - V_1)$

Explanation

Answer. D: The answer is $\frac{F}{A} \times (V_2 - V_1)$. Remember that the work for a piston expanding is pressure multiplied by change in volume: $W = P \times \Delta V$. Because pressure is equal to force over an area, $P = \frac{F}{A}$, and change in volume is $V_2 - V_1$, the resulting equation for the work done is $\frac{F}{A} \times (V_2 - V_1)$.

Fluids

In addition to the behavior of solid particles acted on by forces, it is important to understand the behavior of fluids. First, consider a block of ice, which is solid water. If it is set down inside a large box it will exert a force on the bottom of the box due to its weight as shown on the left, in Part A of the figure below. The solid block exerts a pressure on the bottom of the box equal to its total weight divided by the area of its base:

$$Pressure = Weight\ of\ block/Area\ of\ base$$

That pressure acts only in the area directly under the block of ice.

If the same mass of ice is melted, it behaves much differently. It still has the same weight as before because its mass hasn't changed. However, the volume has decreased because liquid water molecules are more tightly packed together than ice molecules, which is why ice floats (it is less dense).

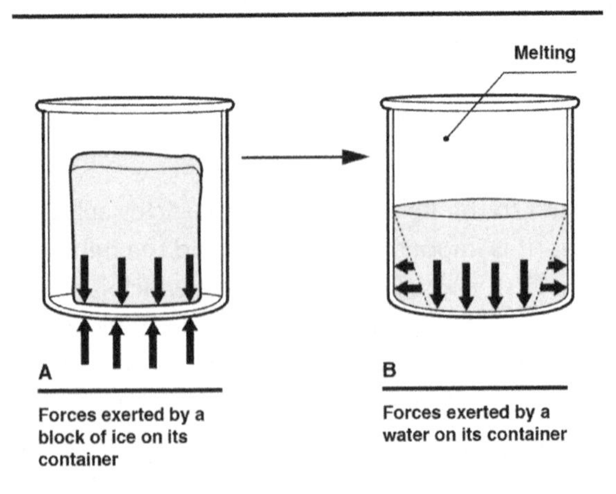

The melted ice conforms to the shape of the container. This means that the fluid exerts pressure not only on the base, but on the sides of the box at the water line and below.

The fact that the liquid exerts pressure in all directions is part of the reason some solids float in liquids. Consider the forces acting on a block of wood floating in water, as shown in the figure below.

Floatation of a Block of Wood

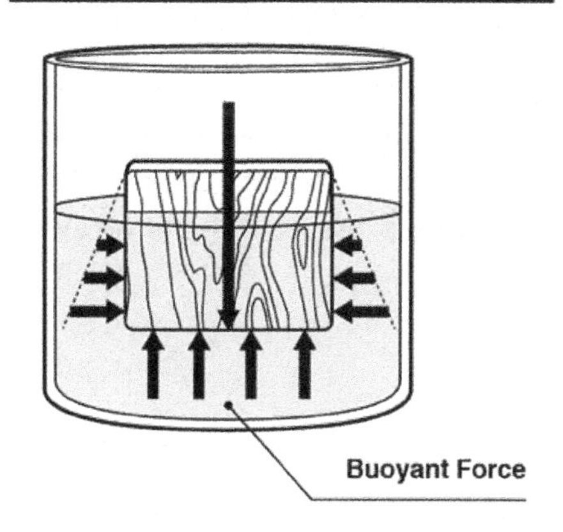

The block of wood is submerged in the water and pressure acts on its bottom and sides as shown. The weight of the block tends to force it down into the water. The force of the pressure on the left side of the block just cancels the force of the pressure on the right side.

There is a net upward force on the bottom of the block due to the pressure of the water acting on that surface. This force, which counteracts the weight of the block, is known as the **buoyant force**.

Pascal's law states that a change in pressure, applied to an enclosed fluid, is transmitted undiminished to every portion of the fluid and to the walls of its containing vessel. This principle is used in the design of hydraulic jacks, as shown in the figure below.

A force (F_1) is exerted on a small "driving" piston, which creates pressure on the hydraulic fluid. This pressure is transmitted through the fluid to a large cylinder. While the pressure is the same everywhere in the oil, the pressure action on the area of the larger cylinder creates a much higher upward force (F_2).

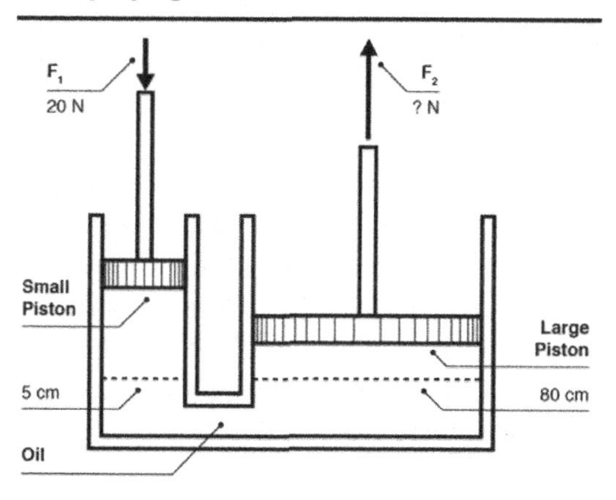

Looking again at the figure above, suppose the diameter of the small cylinder is 5 centimeters and the diameter of the large cylinder is 80 centimeters. If a force of 20 newtons (*N*) is exerted on the small driving piston, what's the value of the upward force F_2? In other words, what weight can the large piston support?

The pressure within the system is created from the force F_1 acting over the area of the piston:

$$P = \frac{F_1}{A} = \frac{20\ N}{\pi\ (0.05\ m)^2/4} = 10{,}185\ Pa$$

The same pressure acts on the larger piston, creating the upward force, F_2:

$$F_2 = P \times A = 10{,}185\ Pa \times \pi \times (0.8\ m)^2/4$$
$$= 5120\ N$$

Because a liquid has no internal shear strength, it can be transported in a pipe or channel between two locations. A fluid's **rate of flow** is the volume of fluid that passes a given location in a given amount of time and is expressed in *m^3/second*. The **flow rate** (*Q*) is determined by measuring the **area of flow** (*A*) in *m^2*, and the **flow velocity** (*v*) in *m/s*:

$$Q = v \times A$$

This equation is called the **Continuity Equation**. It's one of the most important equations in engineering and should be memorized. For example, what is the flow rate for a pipe with an inside diameter of 1200 millimeters running full with a velocity of 1.6 m/s (measured by a **sonic velocity meter**)?

Using the Continuity Equation, the flow is obtained by keeping careful track of units:

$$Q = v \times A = 1.6\frac{m}{s} \times \frac{\pi}{4} \times \left(\frac{1200\,mm}{1000\,mm/m}\right)^2$$
$$= 1.81\,m^3/second$$

For more practice, imagine that a pipe is filling a storage tank with a diameter of 100 meters. How long does it take for the water level to rise by 2 meters?

Since the flow rate (*Q*) is expressed in m^3/second, and volume is measured in m^3, then the time in seconds to supply a volume (*V*) is *V/Q*. Here, the volume required is:

$$Volume\ Required = Base\ Area \times Depth$$
$$= \frac{\pi}{4}100^2 \times 2\,m = 15{,}700\,m^3$$

Thus, the time to fill the tank another 2 meters is 15,700 m^3 divided by 1.81 m^3/s = 8674 seconds or 2.4 hours.

It's important to understand that, for a given flow rate, a smaller pipe requires a higher velocity.

The energy of a flow system is evaluated in terms of potential and kinetic energy, the same way the energy of a falling weight is evaluated. The total energy of a fluid flow system is divided into potential energy of elevation, and pressure and the kinetic energy of velocity. **Bernoulli's Equation** states that, for a constant flow rate, the total energy of the system (divided into components of elevation, pressure, and velocity) remains constant. This is written as:

$$Z + \frac{P}{\rho g} + \frac{v^2}{2g} = Constant$$

Each of the terms in this equation has dimensions of meters. The first term is the **elevation energy**, where Z is the elevation in meters. The second term is the **pressure energy**, where P is the pressure, ρ is the density, and g is the acceleration of gravity. The dimensions of the second term are also in meters. The third term is the **velocity energy**, also expressed in meters.

For a fixed elevation, the equation shows that, as the pressure increases, the velocity decreases. In the other case, as the velocity increases, the pressure decreases.

The use of the Bernoulli Equation is illustrated in the figure below. The total energy is the same at Sections 1 and 2. The area of flow at Section 1 is greater than the area at Section 2. Since the flow rate is the same at each section, the velocity at Point 2 is higher than at Point 1:

$$Q = V_1 \times A_1 = V_2 \times A_2, \qquad V_2 = V_1 \times \frac{A_1}{A_2}$$

Finally, since the total energy is the same at the two sections, the pressure at Point 2 is less than at Point 1. The tubes drawn at Points 1 and 2 would actually have the water levels shown in the figure; the pressure at each point would support a column of water of a height equal to the pressure divided by the unit weight of the water ($h = P/\rho g$).

An Example of Using the Bernoulli Equation

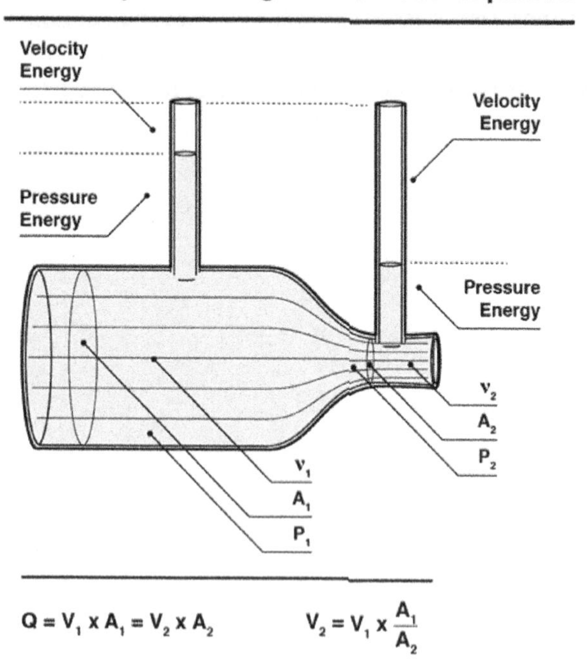

Example

Q. Closed Basins A and B each contain a 10,000-ton block of ice. The ice block in Basin A is floating in sea water. The ice block in Basin B is aground on a rock ledge (as shown). When all the ice melts, what happens to the water level in Basin A and Basin B?

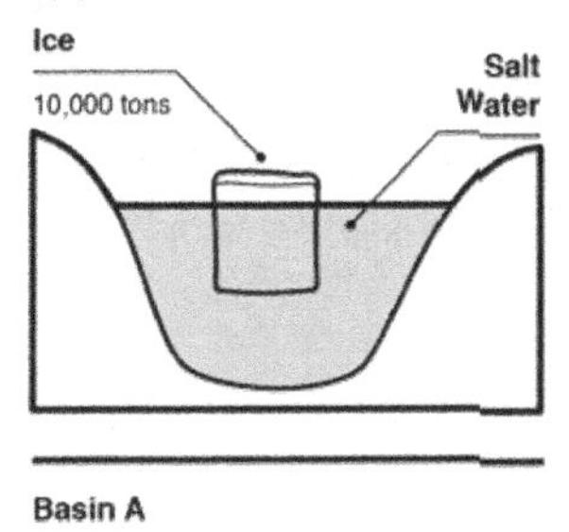

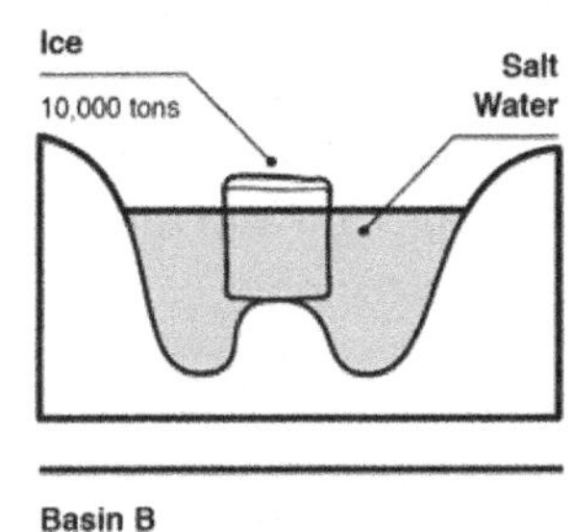

a. Water level rises in A but not in B
b. Water level rises in B but not in A
c. Water level rises in neither A nor B
d. Water level rises in both A and B

Explanation

Answer. B: The answer is that the water level rises in B but not in A. Why? Because ice is not as dense as water, so a given mass of water has more volume in a solid state than in a liquid state. Thus, it floats. As the mass of ice in Basin A melts, its volume (as a liquid) is reduced. In the end, the water level doesn't change. The ice in Basin B isn't floating. It's perched on high ground in the center of the basin. When it melts, water is added to the basin and the water level rises.

Levers

The simplest machine is a **lever**, which consists of two pieces or components: a **bar** (or beam) and a **fulcrum** (the pivot-point around which motion takes place). As shown below, the **effort** acts at a distance (L_1) from the fulcrum and the **load** acts at a distance (L_2) from the fulcrum.

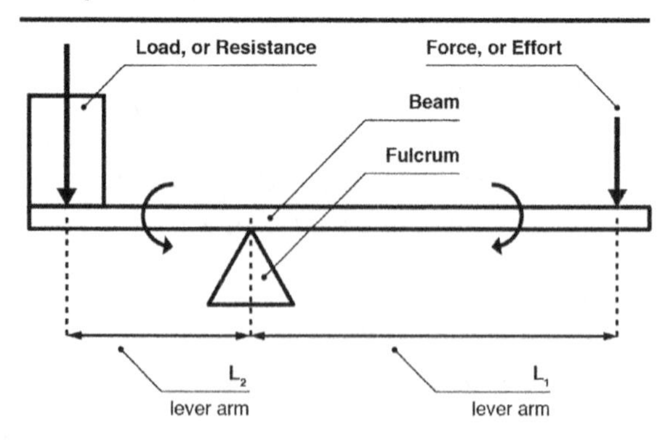

These lengths L_1 and L_2 are called **lever arms**. When the lever is balanced, the load (R) times its lever arm

(L_2) equals the effort (F) times its lever arm (L_1). The force needed to lift the load is:

$$F = R \times \frac{L_2}{L_1}$$

This equation shows that as the lever arm L_1 is increased, the force required to lift the resisting load (R) is reduced. This is why Archimedes, one of the leading ancient Greek scientists, said, "Give me a lever long enough, and a place to stand, and I can move the Earth."

The ratio of the moment arms is the so-called "mechanical advantage" of the simple lever; the effort is multiplied by the mechanical advantage. For example, a 100-kilogram mass (a weight of approximately 1000 N) is lifted with a lever like the one in the figure below, with a total length of 3 meters, and the fulcrum situated 50 centimeters from the left end. What's the force needed to balance the load?

$$F = 1000\ N \times \frac{0.5\ meters}{2.5\ meters} = 200\ N$$

Depending on the location of the load and effort with respect to the fulcrum, three "classes" of lever are recognized. In each case, the forces can be analyzed as described above.

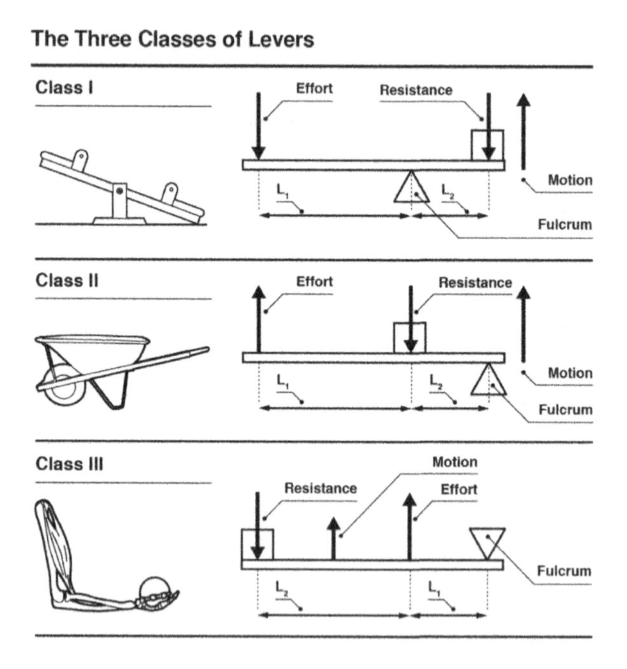

As seen in the figure, a **Class I** lever has the fulcrum positioned between the effort and the load. Examples of Class I levers include see-saws, balance scales, crow bars, and scissors. As explained above, the force needed to balance the load is $F = R \times (L_2/L_1)$, which means that the mechanical advantage is L_2/L_1. The crane boom shown back in the first figure in this section was a Class I lever, where the tower acted as the fulcrum and the counterweight on the left end of the boom provided the effort.

For a **Class II** lever, the load is placed between the fulcrum and the effort. A wheel barrow is a good example of a Class II lever. The mechanical advantage of a Class II lever is $(L_1 + L_2)/L_2$.

For a **Class III** lever, the effort is applied at a point between the fulcrum and the load, which increases the speed at which the load is moved. A human arm is a Class III lever, with the elbow acting as the fulcrum. The mechanical advantage of a Class III lever is $(L_1 + L_2)/L_1$.

Example

Q. A 150-kilogram mass is placed on the left side of the lever as shown. What force must be exerted on the right side (in the location shown by the arrow) to balance the weight of this mass?

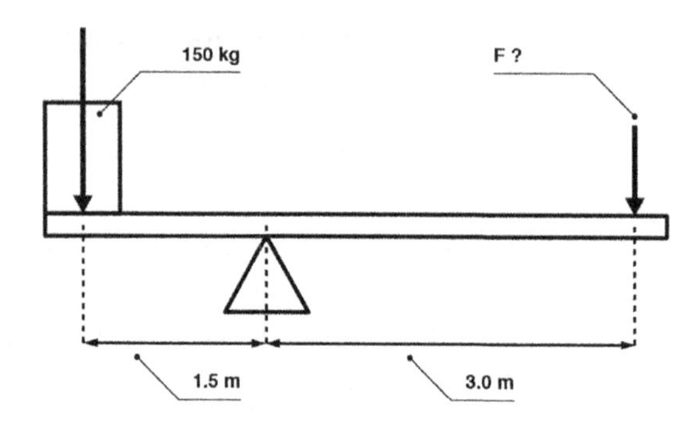

a. 675 kg·m
b. 735.75 N
c. 1471.5 N
d. 2207.25 N·m

Explanation

Answer. B: The answer is 735.75 N. This is a simple calculation:

$$\frac{9.81 \text{ m}}{\text{s}^2} \times 150 \text{ kg} \times 1.5 \text{ m} = 3 \text{ m} \times F$$

Rearranging the equation to solve for the unknown force, F, yields:

$$F = \frac{9.81\frac{\text{m}}{\text{s}^2} \times 150\text{ kg} \times 1.5\text{m}}{3\text{ m}} = \frac{2207.25\text{ N}\cdot\text{m}}{3\text{ m}}$$
$$= 735.75\text{ N}$$

Wheels and Axles

The wheel and axle is a special kind of lever. The **axle**, to which the load is applied, is set perpendicular to the **wheel** through its center. Effort is then applied along the rim of the wheel, either with a cable running around the perimeter or with a **crank** set parallel to the axle.

The mechanical advantage of the wheel and axle is provided by the moment arm of the perimeter cable or crank. Using the center of the axle (with a radius of *r*) as the fulcrum, the resistance of the load (*L*) is just balanced by the effort (*F*) times the wheel radius:

$$F \times R = L \times r \quad \text{or} \quad F = L \times \frac{r}{R}$$

This equation shows that increasing the wheel's radius for a given shaft reduces the required effort to carry the load. Of course, the axle must be made of a strong material or it'll be twisted apart by the applied torque. This is why steel axles are used.

Example

Q. For the wheel and axle assembly shown, the shaft radius is 20 millimeters and the wheel radius is 300 millimeters. What's the required effort to lift a 600 N load?

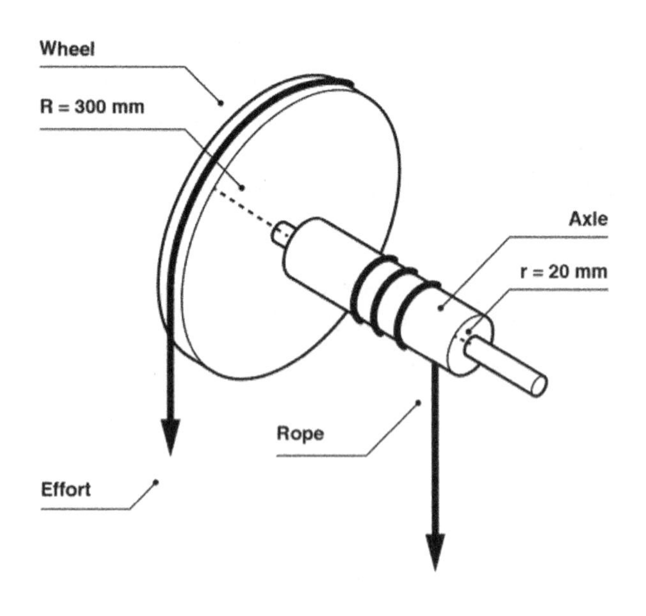

a. 10 N
b. 20 N
c. 30 N
d. 40 N

Explanation
Answer. D: The answer is 40 N. Use the equation $F = L \times r/R$. Note that for an axle with a given, set radius, the larger the radius of the wheel, the greater the mechanical advantage:

$$F = 600\text{ N} \times \frac{20\text{ mm}}{300\text{ mm}} = 40\text{ N}$$

Gears, Belts, and Cams

The functioning of a wheel and axle can be modified with the use of gears and belts. **Gears** are used to change the direction or speed of a wheel's motion.

The direction of a wheel's motion can be changed by using **beveled gears**, with the shafts set at right angles to each other, as shown in part *A* in the figure below.

The speed of a wheel can be changed by meshing together **spur gears** with different diameters. A small gear (A) is shown driving a larger gear (B) in the middle section *(B)* in the figure below. The gears rotate in opposite directions; if the driver, Gear A, moves clockwise, then Gear B is driven counterclockwise. Gear B rotates at half the speed of the driver, Gear A. In general, the change in speed is given by the ratio of the number of teeth in each gear:

$$\frac{Rev_{Gear\ B}}{Rev_{Gear\ A}} = \frac{Number\ of\ Teeth\ in\ A}{Number\ of\ Teeth\ in\ B}$$

Rather than meshing the gears, **belts** are used to connect them as shown in part *(C)*.

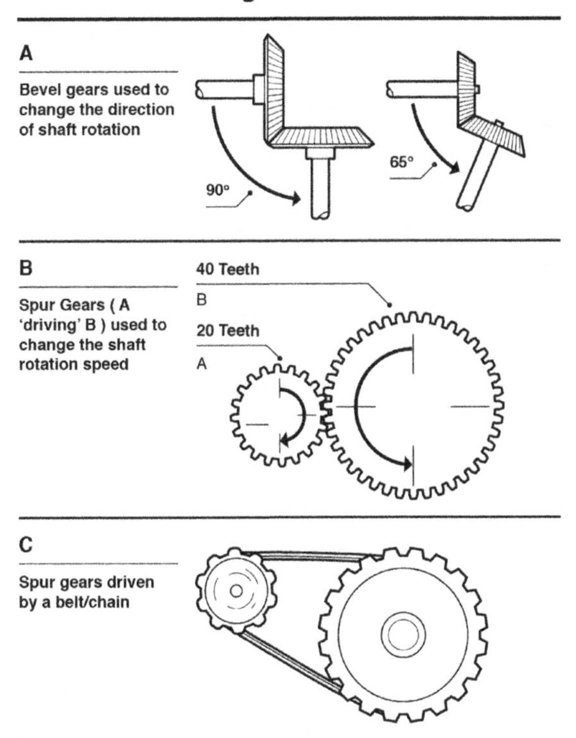

Gears can change the speed and direction of the axle rotation, but the rotary motion is maintained. To convert the rotary motion of a gear train into linear motion, it's necessary to use a **cam** (a type of off-centered wheel shown in the figure below, where rotary shaft motion lifts the valve in a vertical direction.

Conversion of Rotary to Vertical Linear Motion with a Cam

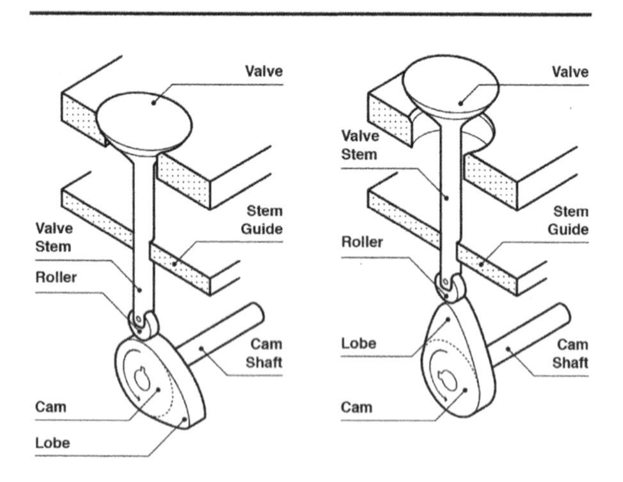

Example

Q. The driver gear (Gear A) turns clockwise at a rate of 60 RPM. In what direction does Gear B turn and at what rotational speed?

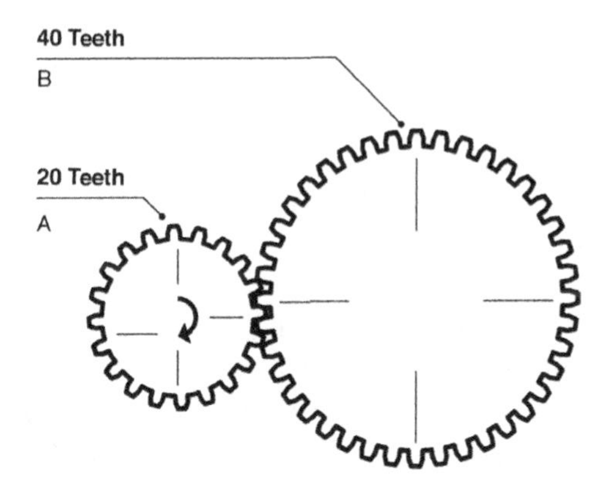

a. Clockwise at 120 RPM
b. Counterclockwise at 120 RPM
c. Clockwise at 30 RPM
d. Counterclockwise at 30 RPM

Explanation

Answer. D: The answer is counterclockwise at 30 RPM. The driver gear is turning clockwise, and the gear meshed with it turns counter to it. Because of the 2:1 gear ratio, every revolution of the driver gear causes half a revolution of the follower.

Pulleys

A **pulley** looks like a wheel and axle but provides a mechanical advantage in a different way. A **fixed pulley** was shown previously as a way to capture the potential energy of a falling weight to do useful work by lifting another weight. As shown in part *A* in the figure below, the fixed pulley is used to change the direction of the downward force exerted by the falling weight, but it doesn't provide any mechanical advantage.

The lever arm of the falling weight (A) is the distance between the rim of the fixed pulley and the center of the axle. This is also the length of the lever arm acting on the rising weight (B), so the ratio of the two arms is 1:0, meaning there's no mechanical advantage. In the case of a wheel and axle, the mechanical advantage is the ratio of the wheel radius to the axle radius.

A **moving pulley**, which is really a Class II lever, provides a mechanical advantage of 2:1 as shown below on the right side of the figure *(B)*.

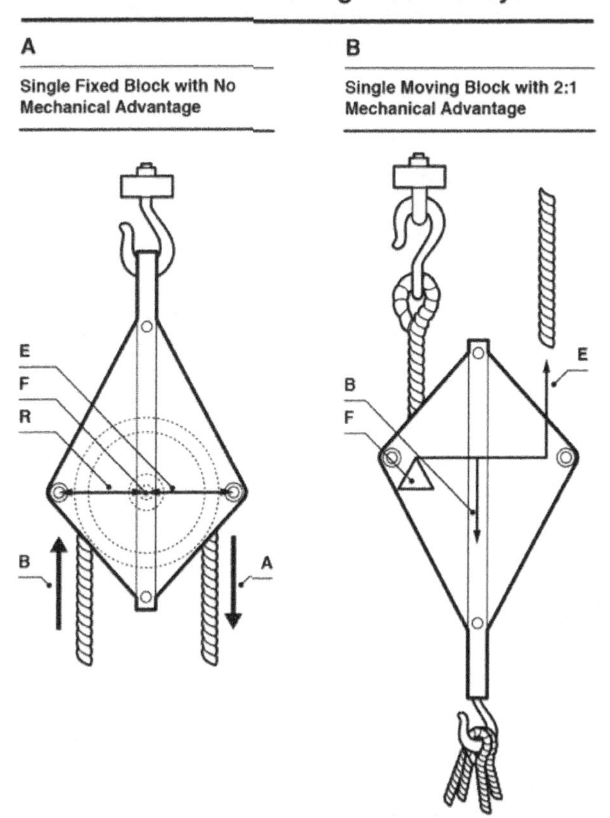

As demonstrated by the rigs in the figure below, using a wider moving block with multiple sheaves can achieve a greater mechanical advantage.

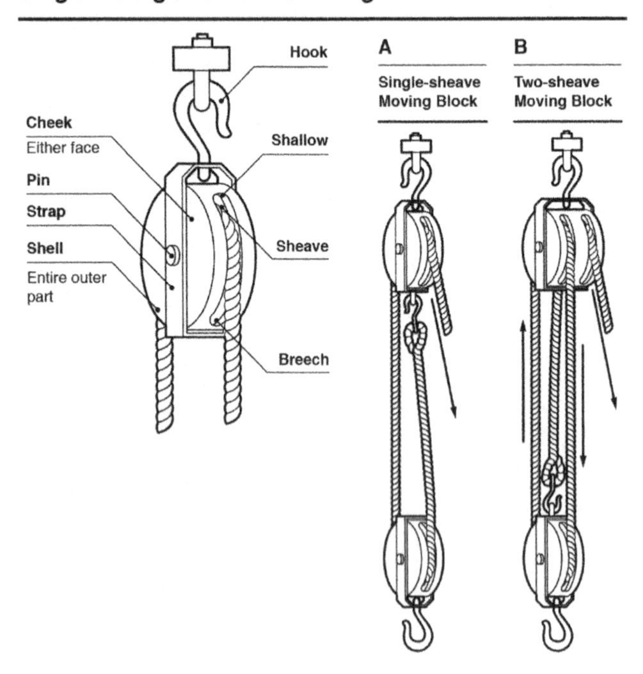

The mechanical advantage of the multiple-sheave block and tackle is approximated by counting the number of ropes going to and from the moving block.

For example, there are two ropes connecting the moving block to the fixed block in part *A* of the figure above, so the mechanical advantage is 2:1. There are three ropes connecting the moving and fixed blocks in part *B*, so the mechanical advantage is 3:1. The advantage of using a multiple-sheave block is the increased hauling power obtained, but there's a cost; the weight of the moving block must be overcome, and a multiple-sheave block is significantly heavier than a single-sheave block.

Example

Q. In case (a), both blocks are fixed. In case (b), the load is hung from a moveable block. Ignoring friction, what is the required force to move the blocks in both cases?

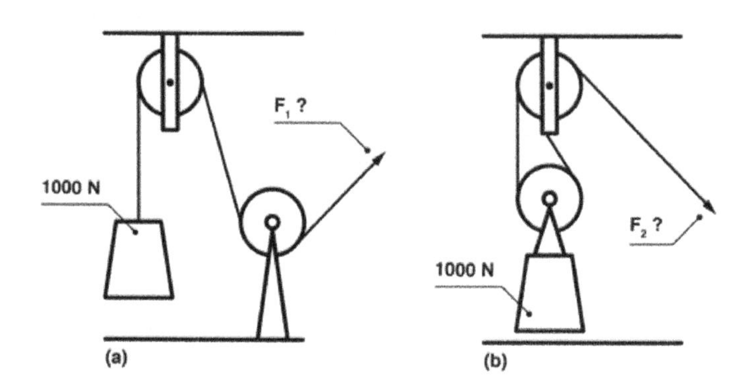

a. $F_1 = 500$ N; $F_2 = 500$ N
b. $F_1 = 500$ N; $F_2 = 1000$ N
c. $F_1 = 1000$ N; $F_2 = 500$ N
d. $F_1 = 1000$ N; $F_2 = 1000$ N

Explanation

Answer. C: The answer is $F_1 = 1000$ N; $F_2 = 500$ N. In case (a), the fixed wheels only serve to change direction. They offer no mechanical advantage because the lever arm on each side of the axle is the same. In case (b), the lower moveable block provides a

2:1 mechanical advantage. A quick method for calculating the mechanical advantage is to count the number of lines supporting the moving block (there are two in this question). Note that there are no moving blocks in case (a).

Ramps

The **ramp** (or inclined plane) has been used since ancient times to move massive, extremely heavy objects up to higher positions, such as in the pyramids of the Middle East and Central America.

Example

For example, to lift a barrel straight up to a height (*H*) requires a force equal to its weight (*W*). However, the force needed to lift the barrel is reduced by rolling it up a ramp, as shown below. So, if the ramp is *D* meters long and *H* meters high, the force (*F*) required to roll the weight (*W*) up the ramp is:

$$F = \frac{H}{D} \times W$$

Definition Sketch for a Ramp or Inclined Plane

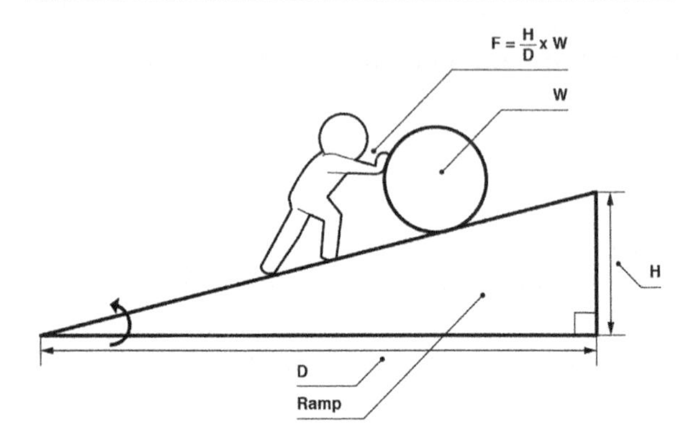

For a fixed height and weight, the longer the ramp, the less force must be applied. Remember, though, that the useful work done (in *N-m*) is the same in either case and is equal to $W \times H$.

Wedges

If an incline or ramp is imagined as a right triangle, then a **wedge** would be formed by placing two inclines (ramps) back to back (or an isosceles triangle). A wedge is one of the six simple machines and is used to cut or split material. It does this by being driven for its full length into the material being cut. This material is then forced apart by a distance equal to the base of the wedge. Axes, chisels, and knives work on the same principle.

Screws

Screws are used in many applications, including vises and jacks. They are also used to fasten wood and other materials together. A screw is thought of as an inclined plane wrapped around a central cylinder. To visualize this, one can think of a barbershop pole, or cutting the shape of an incline (right triangle) out of a sheet of paper and wrapping it around a pencil (as in part *A* in the figure below). Threads are made from steel by turning round stock on a lathe and slowly advancing a cutting tool (a wedge) along it, as shown in part *B*.

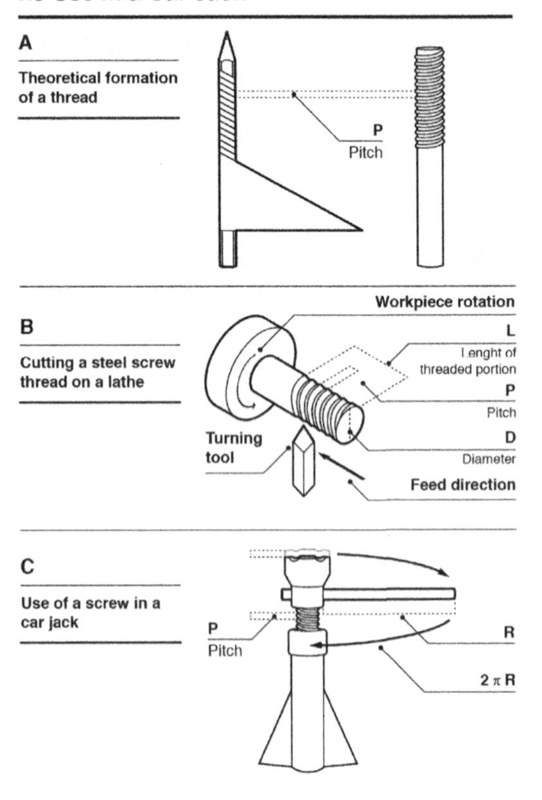

The application of a simple screw in a car jack is shown in part *C* in the figure above. The mechanical advantage of the jack is derived from the pitch of the screw winding. Turning the handle of the jack one revolution raises the screw by a height equal to the **screw pitch** (*p*). If the handle has a length *R*, the distance the handle travels is equal to the circumference of the circle it traces out. The theoretical mechanical advantage of the jack's screw is:

$$MA = \frac{F}{L} = \frac{p}{2\pi R} \text{ so } \quad F = L \times \frac{p}{2\pi R}$$

<u>Example</u>

For example, the theoretical force (*F*) required to lift a car with a mass (*L*) of 5000 kilograms, using a jack with a handle 30 centimeters long and a screw pitch of 0.5 cm, is given as:

$$F \cong 50{,}000\ N \times \frac{0.5\ cm}{6.284 \times 30\ cm} \cong 130\ N$$

The theoretical value of mechanical advantage doesn't account for friction, so the actual force needed to turn the handle is higher than calculated.

Assembling Objects

The **Assembling Objects (AO)** section tests a candidate's ability to think in three dimensions by having them follow a simple set of instructions and assemble an object from its component parts. An example of a question similar to those found on the exam is shown below.

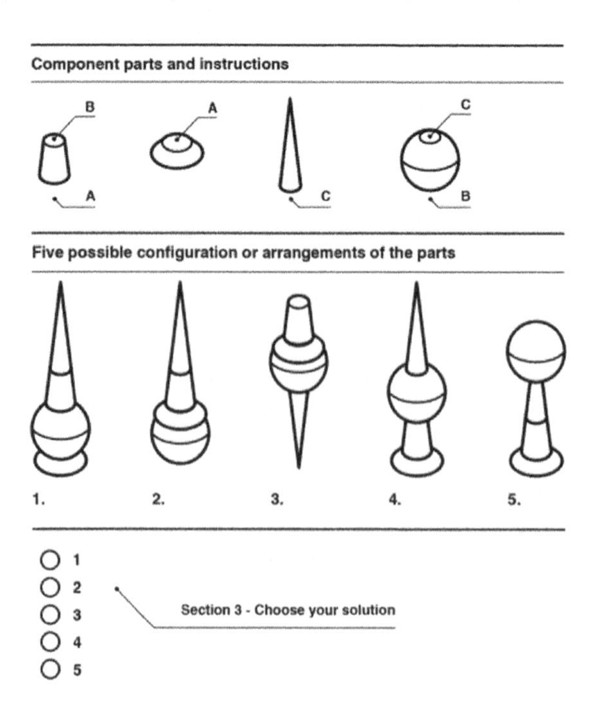

Understanding the Structure of an Assembling Objects (AO) Question

The most efficient way to solve AO questions is to systematically rule out the incorrect answers one at a time. Look at the entire system of parts first, and then work component by component.

In the example above, the system consists of four components: a truncated cone marked *A* on the bottom and *B* on the top face; a shorter, squatter truncated cone labeled *A* on the top face; a thin, tipped cone marked *C* on the base; and a sphere with a line around its "equator" marked *C* on its "north pole" and a *B* on its "south pole." This is a simple system with four components connected "in series," which means one after the other.

Look at Box 1. The thin, tipped cone fits neatly onto the taller of the two truncated cones and sits on top of the sphere. However, the instructions show that the top of the truncated cones must be joined to *B*, and the sharp conic section must be joined at its base to *C*, so these two pieces don't fit together. Therefore, Box 1 is not the solution.

Box 2 can be rejected immediately because it shows the same incorrect assembly.

Now look at Box 3. The two truncated cones are properly connected, with the squat one serving as the base. However, the base is shown sitting on top of the sphere when the surface of a part labeled *C* should be sitting there (which is the sharp cone). In addition, the surface of the squat base is not connected to any part, so Box 3 isn't the answer.

Now that three possible answers have been eliminated, it's easy to compare the last two configurations (Box 4 and Box 5) and see that the only one matching the problem's assembly requirements is Box 4.

Preparing for the Test

The unassembled parts shown in the left-hand box can be moved to different positions, rotated, or even flipped. The parts can also be turned so that only one face is visible. However, the parts won't be stretched or bent, so think of them as being made out of solid wood, rather than out of rubber or plastic. It's also important to remember that the rotations must obey the laws of symmetry. A right-handed object can't be turned so that it becomes a left-handed object. Consider a pair of gloves; a right-hand glove can be turned around or over but, no matter how it's turned or rotated, it can't be turned into a left-hand glove.

Assembling Objects Examples

Section 1: For questions 1–4 below, which figure best shows how the objects in the left box will appear if they are fit together?

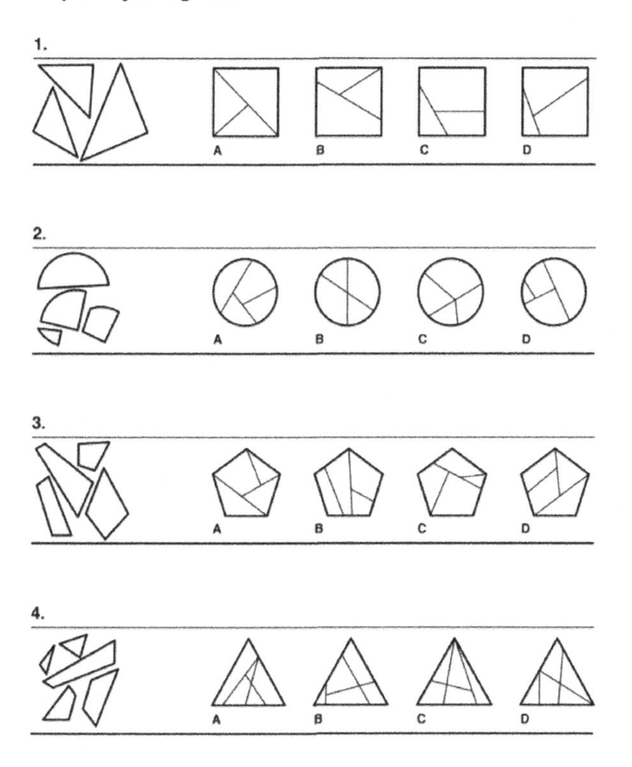

Section 2: For questions 1–3 below, which figure best shows how the objects in the left box will touch if the letters for each object are matched?

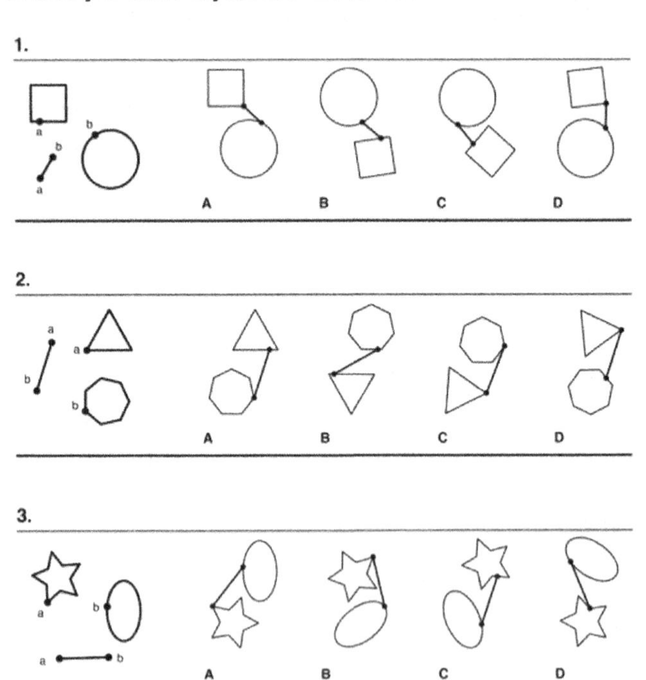

Answers

Section 1

1. A

2. D

3. B

4. A

Section 2

1. B

2. C

3. A

FREE Test Taking Tips DVD Offer

To help us better serve you, we have developed a Test Taking Tips DVD that we would like to give you for FREE. **This DVD covers world-class test taking tips that you can use to be even more successful when you are taking your test.**

All that we ask is that you email us your feedback about your study guide. Please let us know what you thought about it – whether that is good, bad or indifferent.

To get your **FREE Test Taking Tips DVD**, email freedvd@studyguideteam.com with "FREE DVD" in the subject line and the following information in the body of the email:

a. The title of your study guide.

b. Your product rating on a scale of 1-5, with 5 being the highest rating.

c. Your feedback about the study guide. What did you think of it?

d. Your full name and shipping address to send your free DVD.

If you have any questions or concerns, please don't hesitate to contact us at freedvd@studyguideteam.com.

Thanks again!

Made in the USA
Columbia, SC
28 January 2021